Plants, Chemicals and Growth

Plants, Chemicals and Growth

F. C. STEWARD and A. D. KRIKORIAN

Cornell University

State University of New York at Stony Brook

ACADEMIC PRESS New York and London

ACADEMIC PRESS, INC.
111 Fifth Avenue, New York, New York 10003

United Kingdom Edition published by
ACADEMIC PRESS, INC. (LONDON) LTD.
Berkeley Square House, London W1X 6BA

LIBRARY OF CONGRESS CATALOG CARD NUMBER: 76-167780

PRINTED IN THE UNITED STATES OF AMERICA

Contents

CHAPTER 4. The Induction of Growth in Quiescent Cells

CHAPTER 5. Some Growth Regulatory Systems

CHAPTER 6. Growth-Regulating Effects in Free Cell Systems: Morphogenesis

CHAPTER 7. The Range of Biologically Active Compounds

CHAPTER 8. What Do the Growth-Regulating Substances Do?

CHAPTER 9. Concepts and Interpretations of Growth Regulation

CHAPTER 10. Prospects and Problems

Preface

This work was conceived as a concise, technical essay on the general theme covered by its original working title, "Plant Growth Regulators." However, perhaps because our aim was to present a philosophy of the subject rather than yet another compendium describing those chemicals which regulate the growth of plants and detailing the facts of the growth as regulated, the work outgrew its original format. In this, its more extended form, we hope it will be of interest to a wide audience of students, teachers, research workers, and agriculturalists.

Nevertheless, determined attempts were made to restrict its size. We did this because our view is that one will not necessarily learn all about plant growth regulators in general by attempting to know everything about each of them in particular. This is so because it is in the interactions of growth regulators with all other essential features of the responsive system that the truth about them seems to lie.

Therefore, we make no apology for approaching the subject in our own way, for we believe that too much has often been made of the supposed analogy between the plant growth regulators and the hormones of the animal body. Similarly, too little has been made of the special features of the plants, for assuredly they are not animals. No one who studies organisms closely can fail to be impressed by the very different ways in which higher plants and animals are organized and grow. Consequently, contrasts in the way their growth may be regulated (whether endogenously or exogenously) are to be expected and should not be surprising.

Deliberate interventions to modify the way plants grow by the use of chemical agents (whether these are naturally occurring or synthetic) have recently given to biologists new and powerful means to elucidate the many problems of growth and development, even as they have also been exploited for useful, as well as some illicit, ends. For all these reasons, students of biology, or agriculturalists, or even the lay public, should know more about the role of chemical growth regulatory substances in plants; this applies above all to those who advocate their synthesis and application. If this concise treatment contributes to a fresh approach to a most formidable problem, then our efforts will have been rewarded.

This book is based upon the work of many years, supported by research grants from the U. S. Public Health Service (currently GM 09609 from the National Institutes of Health). These grants made the present collaboration possible in the work done at the New York State College of Agriculture at Cornell University, Ithaca, New York. The Research Foundation of the State University of New York assisted Dr. Krikorian in work done at the State University of New York at Stony Brook, Long Island, New York.

F. C. STEWARD
A. D. KRIKORIAN

Introduction

This book is about chemicals and plants; it is also about growth. Since it concerns plants, it should be of interest to people. In a Western industrialized society, man is more remote than ever from problems of seedtime and harvest, of sowing and of growing crops; nevertheless, he is as dependent as ever on plants for food. He is ever more obtrusively dependent on plants for cleansing the air, which is not only "rendered noxious by animals breathing in it" in the normal balance of nature, but is now unnaturally polluted by industrial wastes and the fossil fuels plundered for energy.

In any work about growth, the fundamental biological questions of our time are encountered. No growth originates *de novo,* despite the recurring furor about the imminence of new life to be created in a test tube. Biochemists may skillfully create intricate molecules and substances, as in the synthesis of the substance that is called a gene, but life requires a degree of organization within which such molecules function far beyond any foreseeable fabrication. There is, as yet, no simple chemical alternative to the ability of cells as they grow to make such complexity out of simple random molecules in their environment. Although we may not yet know why cells, organs, and organisms grow, we do know increasingly how they inherit the details of this propensity. We can also readily observe how they grow as they translate the random molecules of earth and air into form and function.

In the "division of labor" among the parts of complex plants, we know that growth is internally regulated by natural "growth regulatory substances." These substances are not part of the plant's makeup, but are set apart for a more regulatory role. We can also intervene in the normal course of growth with a vast array of chemicals never before encountered by plants to change their behavior, control their pests, and greatly modify their yields of products useful to man. As this occurs in an ever more standardized way and in a mass-produced agriculture, one needs constantly to reflect upon the individual cells and organs to know how a single plant responds and, even more so, to trace the active molecules to their ultimate sites of action.

Whether in the miracle of the natural, regulated control of growth and development, or in the menace of the abuse of fertilizers, herbicides, and pesticides, the chemical controls of plant growth are of absorbing interest and of great importance to man.

Plants, Chemicals and Growth

CHAPTER 1 *Some Chemical Regulators: Some Biological Responses*

Many chemical substances are now applied to affect the growth of flowering plants. The angiosperms are the most highly organized, the most conspicuous plants of the earth's cover, and the most important ones to man in agriculture. Some of the substances used are now so familiar that their names have become household words, e.g., 2,4-D, 2,4-dichlorophenoxyacetic acid, and 2,4,5-T, 2,4,5-trichlorophenoxyacetic acid (Audus, 1964; Barth and Mitchell, 1969; King, 1966). In one way or another all these various regulatory substances intervene and exert recognizable effects upon the overall course of plant growth and development, but they do so in a characteristically nonnutrient way. They act not as substances which are built into the "stuff" of which the plants are made, although they may greatly modify the demand which plants make upon their environment for nutrient substances or for water by their effects.

Among the many regulatory substances in use we have controllable and selective herbicides, as well as some substances, e.g., CMU, 3-[*p*-chlorophenyl]-1,1-dimethylurea, and Ammate, ammonium sulfamate, which cause a total kill; we also have substances that stimulate the rooting of cuttings, e.g., indole-3-butyric acid, IBA (see Fig. 1-1), or that cause leaf (Osborne, 1968) or fruit abscission, e.g., Endothal, disodium 3,6-endoxohexahydrophthalate, or that prevent it (see Fig. 1-2); or those associated with fruit setting and artificial parthenocarpy, e.g., α-naphthaleneacetic acid, NAA (see Fig. 1-3), with stem elongation, e.g., Gibrel, gibberellic acid, or with stem retardation (see Fig. 1-4), e.g., Alar, B-9, or 1,1-dimethylaminosuccinamic acid or Cycocel, 2-chloroethyltrimethylammonium chloride (Cathey, 1964); substances affecting the subtle interactions in nature,

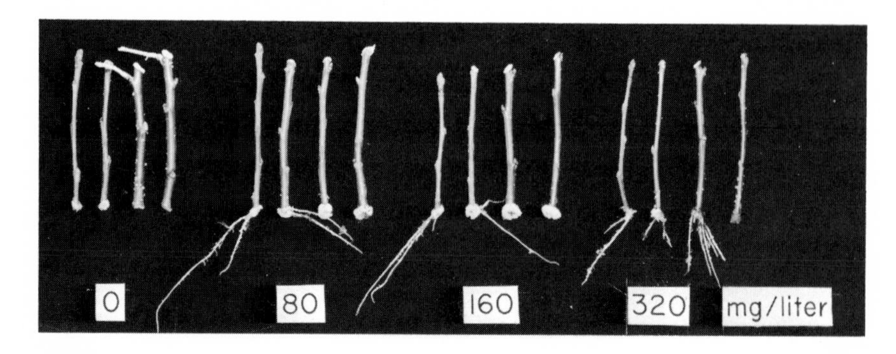

Fig. 1-1. Effects of indolebutyric acid on rooting of apple cuttings after 38 days. In each set of four, the first cuttings retained the shoot tip, the others were taken below the tips. The cuttings were planted in moist peat moss after treatment with the stated concentration for 4 hours. [Photograph supplied by the Boyce Thompson Institute for Plant Research, Inc. (Hitchcock and Zimmerman, 1937/8).]

as between different species (allelopathic substances) (Evenari, 1961; Grümmer, 1961; Muller, 1966; Rice, 1967). There are substances or conditions (temperature, light, etc.) necessary either to promote (e.g., –SH, nitrate) or even inhibit (e.g., coumarin, parasorbic acid, ferulic acid, etc.) germination of seeds in particular instances (Evenari, 1949; Toole *et al.*, 1956), as well as those (e.g., CO_2) to depress (Smock, 1970) or stimulate fruit setting, ripening (e.g., C_2H_4), or respiration (Hansen, 1966). Many useful compilations have been made of these various chemical compounds that have been studied, and of their practical applications (Audus, 1963, 1964; Cathey, 1964; King, 1966; Tukey, 1954).

One can, therefore, approach the problems of growth and growth regulation through the understanding of the roles of chemical substances, natural and synthetic, which intervene to modulate the behavior of the cells of flowering plants, within limits otherwise set by their genetic constitution and after their nutritional demands have been met.

But, inevitably, the observer of growth is impressed with the variety of responses that are involved, such as those to external stimuli and to internally regulated growth correlations. Students of morphology presented adequate descriptions of the regularities and abnormalities of "growth and form" long before their causal explanation could be comprehended in modern terms. One need only look at the transition from the formal descriptive morphology of Goethe (1952) through the attempted causality of Sachs' in terms of the physics and chemistry of his day (Sachs, 1887) which led to the still impressive, experimental approach of Goebel to problems of development [see his *Einleitung in die Experimentelle Morphologie der Pflanzen* (1908)] to see the march of events. But despite the care in ob-

Fig. 1-2. The prevention of apple drop with naphthaleneacetic acid. Left, unsprayed tree; right, sprayed tree. (Photograph courtesy of the U. S. Department of Agriculture.)

servation and the oftentimes definitive descriptions of morphogenetic events in these older works, they lacked the chemical techniques which have produced the present wealth of detailed, but often uncoordinated, observations in the modern period.

A modern student is faced with a seemingly overwhelming array of responses. These include various tropisms (photo- and geotropisms, Curry, 1969; Audus, 1969) and chemotropisms; nastic responses (Ball, 1969); rhythmic phenomena in growth and development (Bünning, 1967; Cumming and Wagner, 1968; Hamner, 1963; Wilkins, 1969); growth of organs by cell division and enlargement; initiation of lateral organs and problems of phyllotaxy (Richards and Schwabe, 1969); the induction of flowering (Lang, 1965) and the formation of such vegetative organs of perennation as buds, tubers, bulbs, etc. (Gregory, 1965; Vegis, 1964; Wareing, 1969a); periodicities in growth as in cambial activity; the regulatory effects of light (Hillman, 1967, 1969; Mohr, 1969) and temperature on growth and form (Chouard, 1960; Hartsema, 1961; Picard, 1968); and the many factors that impinge upon the balance between vegetative growth, flowering (Evans, 1969a), sexuality (see Dzhaparidze, 1967; Heslop-Harrison, 1956, 1964), and fruiting, followed by the formation, dormancy, and germination of seeds (Mayer and Poljakoff-Mayber, 1963; Nikolaeva, 1969). Over and above the problems presented by the more usual forms of develop-

Fig. 1-3. The induction of parthenocarpic strawberry fruits by naphthaleneacetic acid. Left, normal control; center, strawberry which had its achenes removed, treated with lanolin paste; right, strawberry which had its achenes removed, treated with lanolin paste plus naphthaleneacetic acid (100 ppm). [Photograph supplied by Dr. J. P. Nitsch, C.N.R.S., Gif-sur-Yvette (Nitsch, 1950).]

ment enumerated above, there are others presented by abnormal or unusual ones such as the root nodules associated with symbiotic nitrogen fixation (Raggio and Raggio, 1962); the unusual growth of shoot apices which may lead to regular but anomalous phyllotaxy, and the equally anomalous growth of some flowers (for example, of bulb plants) in response to inappropriate exposures to temperature (Hartsema, 1961; Luyten *et al.*, 1926) (Fig. 1-5); and finally such pathological expressions of growth as those seen in teratomas, galls, "witches brooms," and tumors, whether induced by bacteria, insects, mycoplasma, viruses, or even those due to genetic imbalances, as in certain hybrids (see Braun, 1969a). Chemical causation is implicit in all these morphogenetic responses (see Fig. 1-6).

Faced with all this, the trend has been to invent classes of substances that correspond with biological responses, and we now have a plethora of these which include the historically important auxins, the gibberellins, the cytokinins, florigen(s) and anthesin(s), vernalin(s), dormin(s), or abscisin(s), and the recent and even more diffuse class of morphactins. Added to the already formidable array of naturally occurring growth substances are the innumerable products of synthetic chemistry which may also intervene to regulate the growth and behavior of plants. All this has emerged

Fig. 1-4. The retardation of stem length in *Chrysanthemum* by B-9 or Alar. Left, control plants; center, plants sprayed with 2500 ppm B-9; right, plants sprayed with 5000 ppm B-9. (Photograph courtesy of the U. S. Department of Agriculture.)

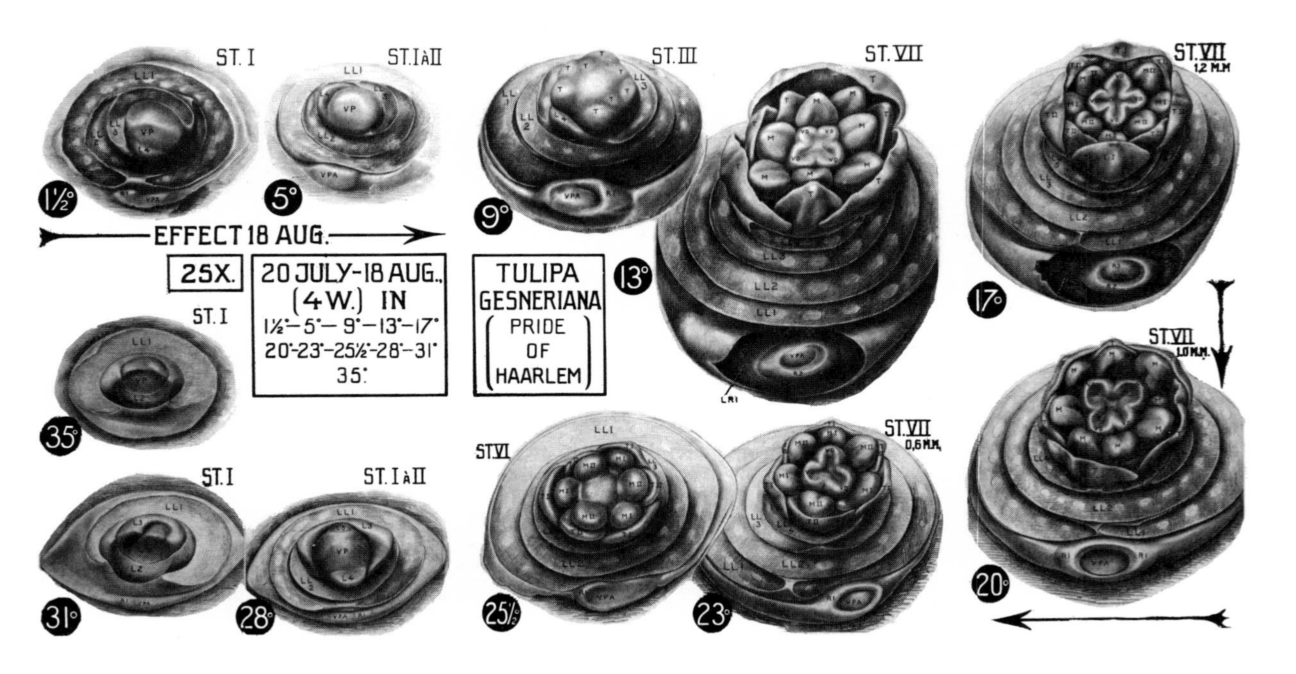

Fig. 1-5. Effects of temperature on the development of flowers in *Tulipa*. The figure shows the apex after 4 weeks of storage at temperatures from 1.5 to 35°C. Normal flowers developed at 23°–25°C; abnormal tetramerous flowers were initiated at 9°–13°C; at 17° to 20°C intermediate numbers of flower parts were formed; at low, i.e., 1.5°C and at high, i.e., 31° and 35°C, the growing point remained vegetative; at 5° and 28°C the growing point was about to initiate flowers; at 9°C only petals had been initiated. This series emphasizes the need for a succession of distinct temperature-dependent morphogenetic stimuli. [Photograph supplied by the Agricultural University. Wageningen (Luyten *et al.*, 1926).]

Fig. 1-6. Some anomalous expressions of growth. A, "Witches' broom" on *Tilia*; B, "Cherry gall," due to a cynipid, on veins of *Quercus* leaf; C, galls on bramble (*Rubus*) induced by gall midges; D, root nodules of soybean (*Glycine max*). [Photograph A–C courtesy of M. J. D. Hirons (Darlington, 1968); D, courtesy of the Nitragin Company, Milwaukee.]

from a period of great productivity, but it cannot be denied that the present is also a period of great confusion.

The experimental results to be reviewed in this work will mainly relate to the growth of flowering plants. This choice should not disparage knowledge which has been gained by work upon other forms. The convenience of microorganisms such as bacteria, yeasts, other fungi, algae, bryophytes, and ferns for the study of many aspects of growth and its regulation is self-evident, and there is a large literature on these organisms. While work with the more highly organized flowering plants presents its own special difficulties, it also presents a special challenge for it concerns the growth of the plants which are so important to man and his environment.

CHAPTER 2 *The Totipotency of Cells and Their Exogenous Regulation*

The "Division of Labor"

The possibilities inherent in the chemical regulation of plants need to be related to the way that flowering plants grow and develop. Like any sexually reproduced organism, a flowering plant as a unique individual begins as a single cell, the fertilized egg or zygote (see Fig. 2-1). The zygote has a fixed complement of genetic information in its nucleus and the machinery in its cytoplasm to transcribe it into effect. Without the latter, the former is of no avail. At its formation, the single-celled zygote may be free in the sense that although enclosed within the embryo sac, it is not in organic connection with the parent plant body (Fig. 2-2), but it differs from a cell of a free-living autotrophic green plant, e.g., a unicellular alga, because it is incapable of a completely separate existence in an inorganic world. Before this autotrophic capacity exists, the fertilized egg must develop while it is nourished heterotrophically by an array of complex metabolites, and receives stimuli to encourage it to develop into an embryo (Wardlaw, 1965a). When this point is reached, specialized functions distinguish its organs and a "division of labor" appears (see Fig. 2-3). This division of labor, or the assignment of function to special organs with prescribed forms, occurs without irreversible loss of the essential genetic information that the zygote received or, ultimately, of the ability of the cytoplasm to transcribe and use that information. This is so because, in a satisfying number of cases, previously mature living cells, isolated from mature plant organs or tissues, may behave again like zygotes and grow like embryos (Halperin, 1969; Reinert, 1968; Steward *et al.*, 1969, 1970; Torrey, 1966). Thus, higher plants achieve their diversity of form and function more by arranging their living cells than by changing them in essential or irreversible ways. The specialization of cells in the organs of higher animals appears earlier in development, is more extreme, and is certainly more difficult to reverse (Davidson, 1969; Gurdon, 1968). Although the work of Gurdon shows that the nuclei of cells may retain their genetic competency throughout

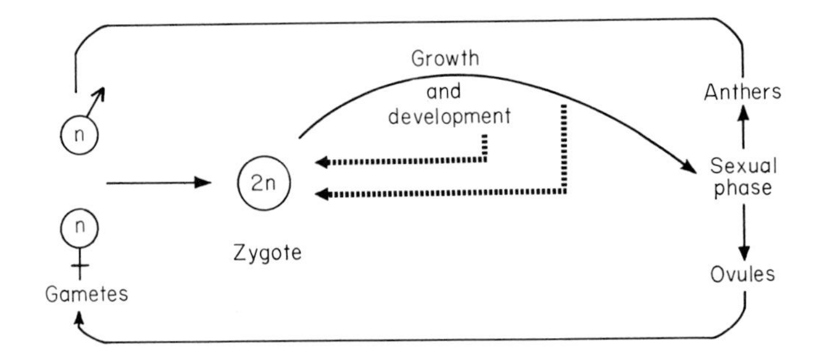

Fig. 2-1. The angiosperm life cycle.

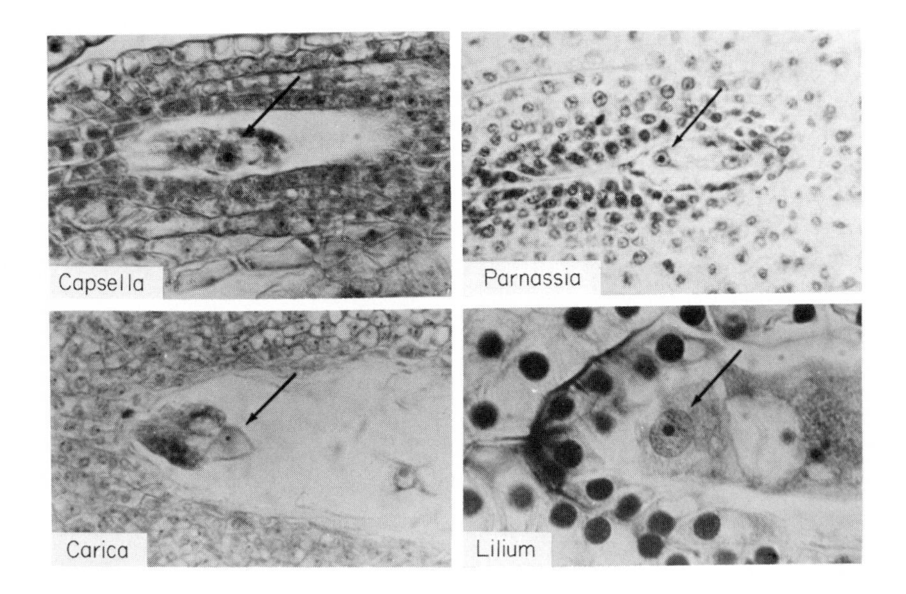

Fig. 2-2. Some representative embryo sacs. Arrows show their zygotes.

the development of a frog, it is yet to be shown that their cytoplasm retains all the properties that were inherent in the egg (Gurdon, 1968).

In sharp contrast to warm-blooded animals, higher plants expose their cells to, and make use of, much wider changes in their immediate environ-

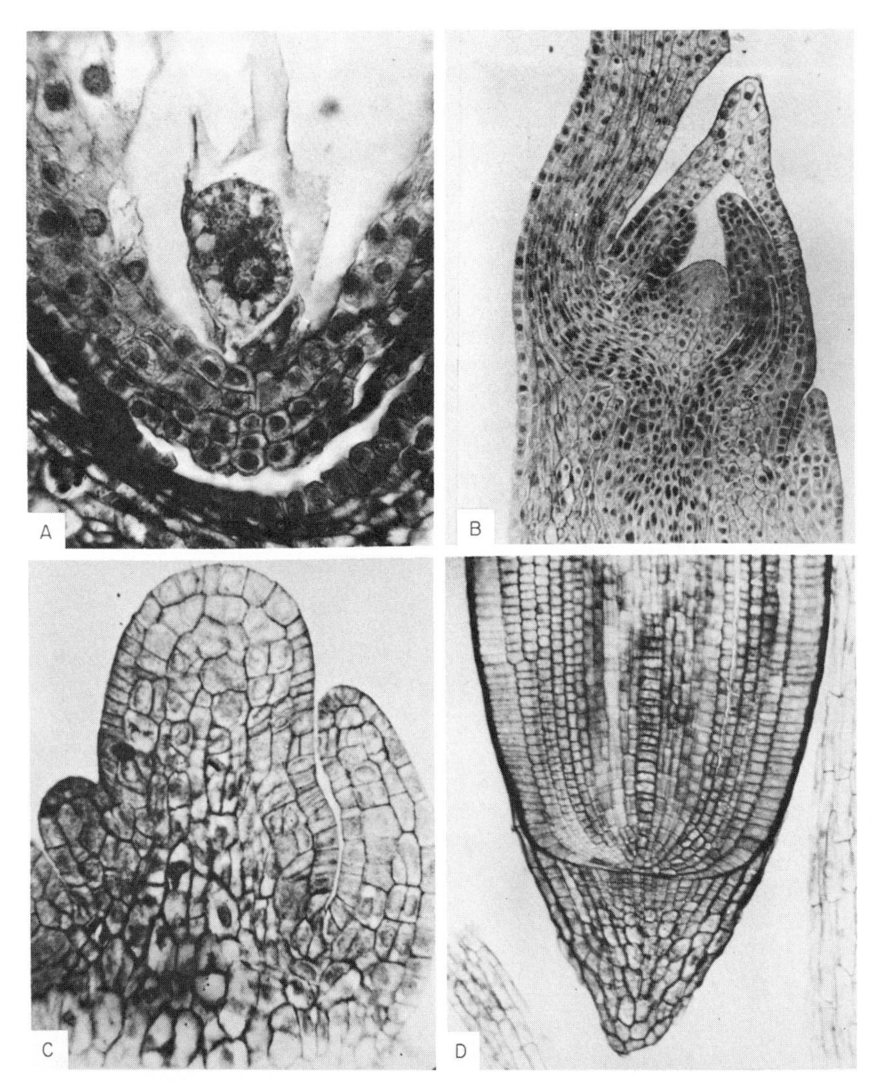

Fig. 2-3. Embryo development and organ formation. A, the proembryo of oats (*Avena*) at the two-celled stage in an embryo sac; B, the shoot apex of an embryo of oats, showing leaf primordia enclosed within the coleoptile; C, apex of the shoot after it has emerged from the coleoptile; D, primary root, showing root cap and zonation in the root apex. [A–D supplied by Dr. O. T. Bonnet, University of Illinois (Bonnet, 1961).]

Fig. 2-4(A). For legend see facing page.

ment (see Fig. 2-4). There is nothing comparable in the plant body to the constancy of internal environment (so-called *homeostasis*), represented by the composition of the blood, by the close control of temperature and water content within very narrow limits, and by the rigorous protective devices against "foreign substances" revealed by antibody–antigen relationships. These facts emphasize that there are many inherent differences and even undisclosed possibilities of chemical growth regulation in plants.

The innate potentialities of any of the living cells of the plant body, which have retained the genetic capacities of their zygotes by equational divisions, nevertheless are modified *in situ* by extrinsic factors. This is traceable to the polarity that develops very early in embryogeny (see Maheshwari, 1950; Wardlaw, 1965a, 1968), i.e., to the complementarity of form and function as between shoot and root, but it is also implicit in the limitations

Fig. 2-4(B)

Fig. 2-4. Effects of environment on development. A, response of *Hyoscyamus niger* (henbane) to photoperiod. Left, plant grown on short days (8 hours); right, plant grown on long days (16 hours); B, response of *Digitalis purpurea* (foxglove) to temperature and darkness. Left, plant grown at normal day length and warm greenhouse temperatures; right, plant grown under cold nights (50° F) and long warm days. [A, photograph courtesy of the U. S. Department of Agriculture; B, photograph supplied by the Boyce Thompson Institute for Plant Research, Inc. (Arthur and Harvill, 1941/2).]

placed upon, or the stimuli received by, those cells of the developing plant body which can act as initiating cells and so form the primordia of developing organs as they grow. Therefore, growth and development are regulated from without and within, by an interchange of "messages" communicated by substances or stimuli that control what the essentially totipotent cells do *in situ*. This scheme of things gives rise to the range of naturally occurring growth regulatory substances, to the responsiveness of the cells to such substances, and to the possibilities inherent in the use of their synthetic analogs, which may intervene from without.

Cell Growth and Cell Division

One should see the processes of growth in distinct and somewhat separable stages. The multiplication of plant cells, by equational division, with the self-duplication of their essential organelles, is the first prominent phase or act. Some prefer to describe this as part of what is interpreted as development. Cell division is normally followed by cell enlargement, an act more typical of plants than of animals. Plant cells increase their internal volume enormously but with proportionately much less increase of cell substance, by "spreading their cytoplasm out thin" so as to maintain the maximum contact with their environment from which, as autotrophs, their essential nutrients are derived. In so doing, vacuoles are created and enlarge as cells acquire and preserve their turgor (see Fig. 2-5). Therefore, cell division and cell enlargement have very different requirements and they also respond to very different stimuli.

Certain criteria need to be established at the outset. For those who work with bacteria or yeast, growth is largely, if not exclusively, synonymous with the multiplication of cells. At the other extreme, and for those whose work begins with preformed seeds, growth may seem synonymous with increase of size. As Fig. 2-6 of the embryo from an oat grain shows, much cell division has already occurred together with organ formation (Avery, 1930). The coleoptile, an organ favored as an experimental object, has usually acquired all its cells before it is isolated for use in growth experiments (see Boysen Jensen, 1936, and references there cited). Therefore, cell enlargement is the aspect of its growth upon which attention is focused. Similar points may be made for the hypocotyl of dicotyledonous plantlets. Some have restricted the use of the term growth to those cases where there is tangible increase of size or substance, and they regard the problems of cell multiplication to be in a different category which is concerned solely

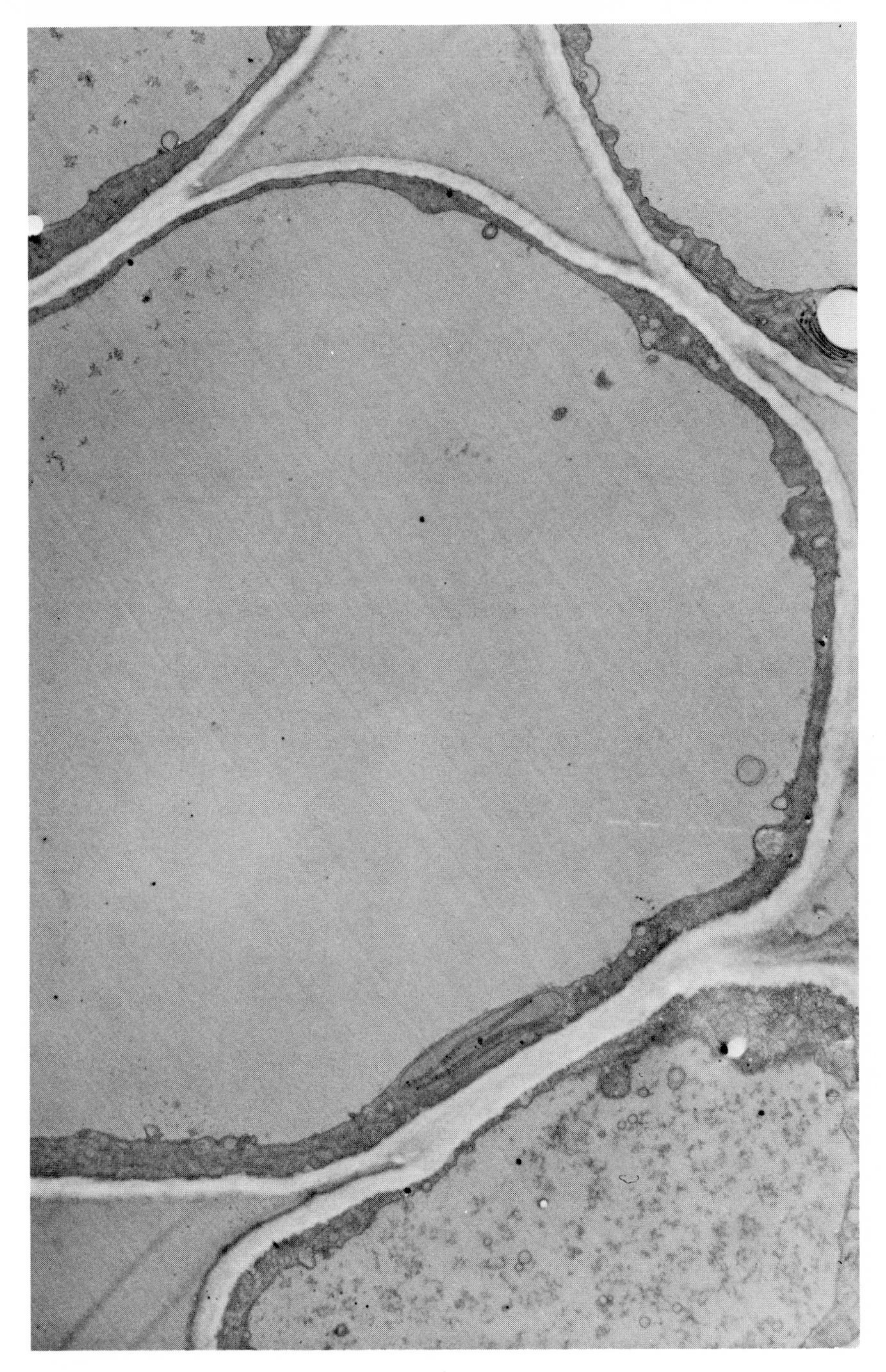

Fig. 2-5. Quiescent cells from the secondary phloem of the carrot root to show, relatively, their cell walls with air spaces, very thin parietal protoplasm, and very large central vacuoles. × 6000. (Preparation by Dr. H. W. Israel.)

with cell replication or the duplication of self-duplicating units. In one definition, Thimann wrote (see Thimann, 1948, 1952, p. 4): "An auxin is an organic substance which promotes growth (i.e. irreversible increase in volume) along the longitudinal axis when applied in low concentration to shoots of plants freed as far as practical from their own inherent growth promoting substance." Because of the inherent difficulties in the application of this definition which followed upon the discoveries of gibberellins, Thimann later (1969) modified the definition of an auxin by adding the phrase "and inhibit the elongation of roots." He further commented that a growth hormone must "of necessity cause cell enlargement," and stated "It could

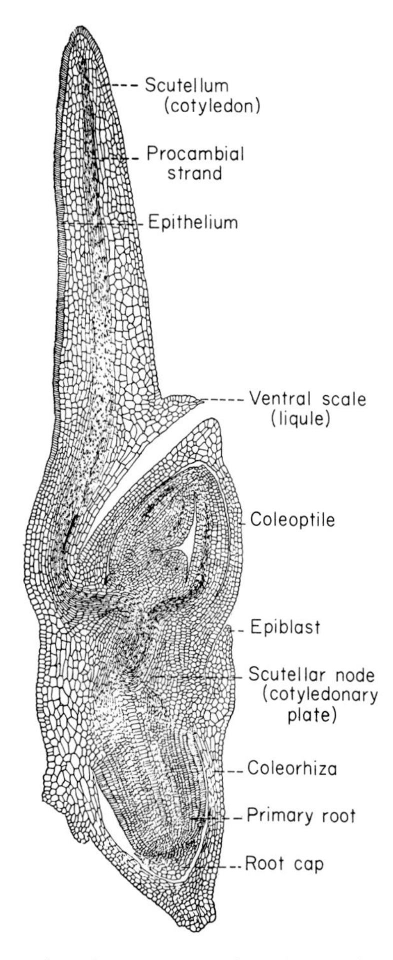

Fig. 2-6. The embryo from an ungerminated oat grain (Avery, 1930).

be wished that biologists would not continue to speak of 'growth by cell division' for the division of cells does not of itself *cause* enlargement, growth is always *by* cell enlargement" (see Thimann, 1969, p. 16). Growth is clearly made synonymous with increase of volume in these quotations. From the standpoint of this work, this is an extreme point of view which is inapplicable.

The point of view adopted here is that growth requires an appropriate balance between cell multiplication and cell enlargement. One cannot conceive of the growth of an organ in a multicellular organism occurring solely by cell enlargement alone, for the cells must first be laid down in the primordia of the organs in question. Nevertheless, there are many situations in which, before actual growth measurements are made, virtually all the cell divisions are completed and the increase that is actu-

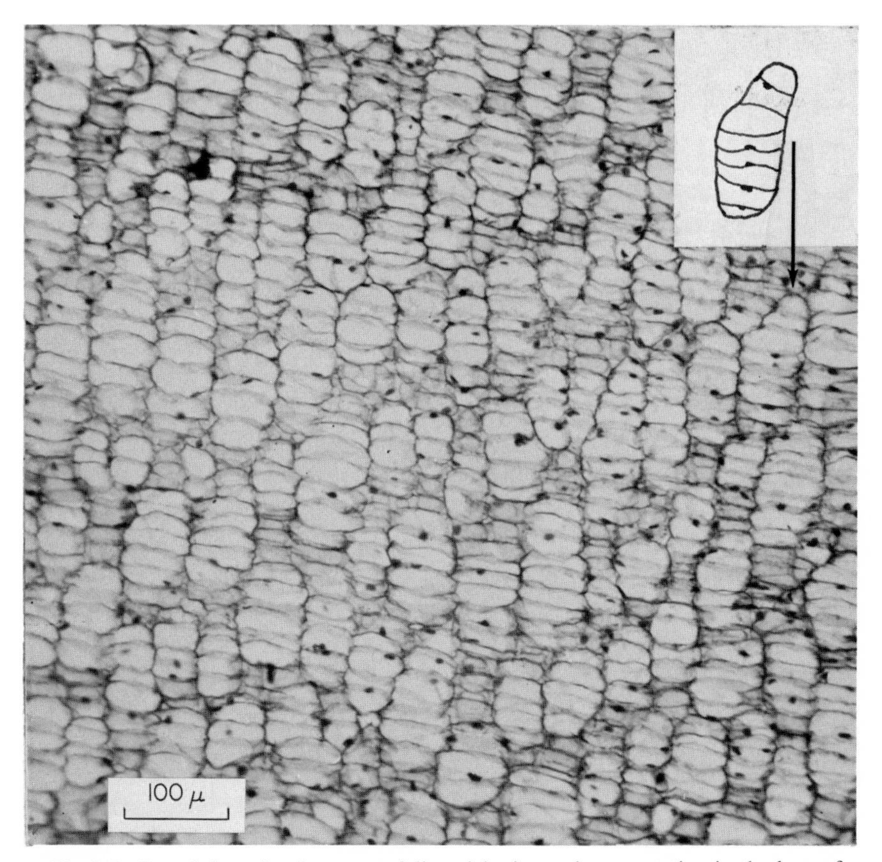

Fig. 2-7. Growth by cell enlargement followed by internal segmentation in the base of a banana leaf (Barker and Steward, 1962).

ally measured occurs solely by cell enlargement. It will be convenient, therefore, to divide the overall problem of growth of angiosperm cells and organs into a phase of (1) cell multiplication and (2) cell enlargement even though these will often overlap. Of course, a certain amount of cell enlargement is often implicit even when growth occurs predominantly by cell division. There are also many examples in which cells first enlarge and then divide internally (see Fig. 2-7). The latter event, i.e., internal segmentation, is as much an aspect of growth as the first, especially if the segments so formed later enlarge.

Unquestionably, the dramatic event which occurs in the growth of higher plants is the overall increase in the number of their cells. If one examines the size of cells in comparable parts of the plant body of a seedling and of a mature plant, one finds that the actual sizes of cells in homologous tissues of the seedling and the adult are remarkably similar. Hence, it is correct to give the multiplication of cells a cardinal place in any discussion of growth in flowering plants although, paradoxically, the literature of plant physiology has given greater prominence, historically, to those aspects which concern cell enlargement. The reason is that the latter are more easily perceived and have proved simpler to attack experimentally. Modern cell biology, however, has made it possible to deal precisely with the sequential events of cell division (Padilla *et al.*, 1969; Prescott, 1964).

The Cell Cycle

The cyclical events of cell multiplication involve both morphological events and biochemical syntheses; the replication of visible structures (John and Lewis, 1969) and organelles, as well as of macromolecules (Stern, 1966). Classical cytology of the nucleus and the wealth of modern knowledge that deals with cells in division and in interphase comprises a literature too vast to be comprehended here. Obviously, such considerations are fundamental to any understanding of the different ways by which chemical substances may intervene to affect cell growth and cell division (Gifford and Nitsch, 1969) to know how and when they act in relation to the cell cycle (Kihlman, 1966).

The impressive feature is the precision with which the morphologically complex events of cell multiplication work in somatic cells as they divide equationally (John and Lewis, 1968). Nevertheless, there are deviations—in ontogeny, as well as due to external stimuli—from the normal plan. The outstanding deviation is of course at meiosis when divisions occur to produce spores with the reduced (haploid) complement of chromosomes. Although reduction divisions occur in anthers and ovules under the demonstrable conditions that control flowering, these conditions have not been translated into the effect of a specific chemical agent which will actually cause the meiotic divisions to occur—important as this knowledge would be (John and Lewis, 1965; Tulecke, 1965). Pollen mother cells divide meiotically to give microspores (n) and these normally give rise to pollen grains, pollen tubes, and male gametes (Stern and Hotta, 1968). However, in selected cases, such pollen cells will divide again in isolated anthers when they are cut open and placed on a culture medium, and the divided haploid cells may even develop into plantlets (Guha and Maheshwari, 1966; Nakata and Tanaka, 1968; Niizeki and Oono, 1968; Nitsch and Nitsch, 1969; Sunderland and Wicks, 1969). The well-known effect of colchicine operates in a converse way to permit cells to retain multiples of their original chromosome complements. Polyteny in chromosomes, conspicuous in insect larvae (Gall, 1963), is rare in angiosperms although it does occur in suspensor cells of some embryos (Nagl, 1969) and, occasionally cells (as in the growing regions of a banana) may even retain two complete nuclei in one cell (Barker and Steward, 1962). Aneuploidy and polyploidy represent deviations from the normal sequence of cell division and their causation and significance for normal development or in the cultivation of explanted angiosperm cells is a large topic in itself (D'Amato, 1952; John and Lewis, 1968; Partanen, 1959). But a distinction should be made between external factors, or substances, that operate during division to affect the morphology of the divided cells (and hence their gene expression) and others which can operate during interphase to affect the controls of biochemistry. Obviously any substance that operates through the synthesis of nucleotides, now attributable to the nucleolus, e.g., as in tRNA's, should exert its effect on interphase, not dividing, nuclei (Birnstiel, 1967).

The modern knowledge which has been gained through the study of fine structure by electron radioautography to locate syntheses now permits one to trace the events of cell growth and division, which comprise the cell cycle, in biochemical terms. The cell cycle is described in terms of those classes of compounds about which most is known and which are most concerned with self-duplication of substances and organelles, i.e., the nucleic acids (DNA and RNA) and the proteins including the histones (Padilla *et al.*, 1969; Prescott, 1964). This is the important background of knowledge against which one should be able to infer when and where many of the chemical plant growth regulators act (Kihlman, 1966; Van't Hof, 1968a,b). The essential steps in the cell cycle, which apply to both plant and animal cells, are summarized in Fig. 2-8.

This figure illustrates the cell cycle in terms of the duration of its distinct

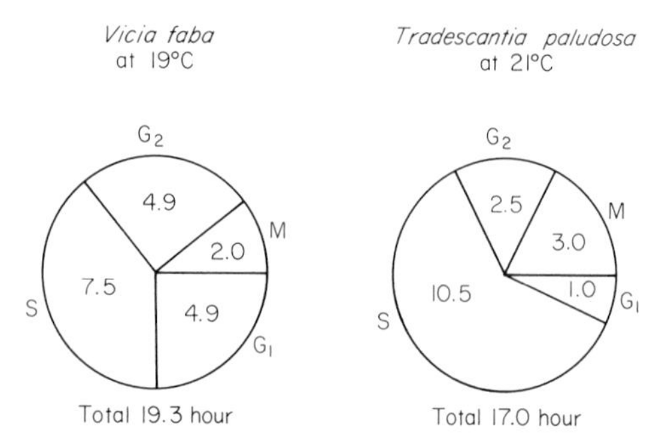

Fig. 2-8. The cell cycle. Drawn from data of Evans and Scott (1964) and Wimber and Quastler (1963). For details see text.

stages with respect to rapidly dividing cells in *Vicia* (Evans and Scott, 1964) and *Tradescantia* root tips (Wimber and Quastler, 1963). The conspicuous visible events of nuclear and cytoplasmic division, i.e., mitosis (M), are *in toto* relatively short. The remaining period comprises those metabolic events which mobilize the necessary substances at their respective sites. These events in sequence are as follows:

G_1 is the interval between mitosis and the onset of nucleic acid (DNA and RNA) synthesis preparatory to the next division. In this substantial period the raw materials for nucleic acid synthesis must be fabricated and mobilized.

S is a principal period of DNA synthesis in which the DNA content per cell is doubled, whereas

G_2 represents a period between the end of the synthesis of DNA and the first visible signs of mitosis. In this period the prefabricated DNA must be organized into the visible structures.

There is very detailed knowledge of the structure of DNA; and in the light of this, of the way in which the DNA molecule replicates (Watson, 1970). Nevertheless, there is virtually no explanation why the DNA, in its milieu in the cell, should replicate (see "Replication of DNA in Micro-Organisms," 1968). This is a way of saying that we do not yet know *why* cells grow, although we know much about *how* they grow. Our ability to understand how growth factors intervene to modify growth in cells is inevitably limited by this state of knowledge.

The above is mainly concerned with the behavior of the DNA. At the two stages at which DNA is involved with either its self-replication, or with partition between daughter nuclei, its role in terms of RNA synthesis is at a minimum, i.e., at S and M; similarly, at the same stages, RNA-mediated protein synthesis is also at a minimum. By contrast, new histone synthesis occurs concomitantly with new DNA synthesis during S (Prescott, 1964; Padilla *et al.*, 1969).

Cellular Ontogeny

Normally, cells which emerge from active cell division, as in a meristem, may embark on a protracted period of cell enlargement, (Wardlaw, 1965b). This involves various syntheses with much less frequent multiplication of cells and, even when it is completed, some cells may enter upon an indefinite period of quiescence during which their metabolism continues and their products and storage of what are often referred to as "secondary products" may occur.

Obviously, in all this, there are anabolic events in which complexity is created, and there are catabolic events in which complexity is destroyed and energy is released. A physiologically active growth regulatory substance could potentially intervene at any one of these crucial phases or points and, ideally, one should know precisely where it acts. But circumstances are known in which development can occur in higher plants without cell multiplication. Seeds exposed to γ-radiation undergo development with no, or minimal, cell divisions, while cell enlargement and maturation continues (Haber and Foard, 1964). In this situation, the normal balance of cell division and enlargement is disturbed (Haber, 1968).

Although cells of unicellular microorganisms divide and separate, higher plants only achieve their complexity and diversity of functions as cells remain attached and in organic connection. The patterns of cell growth and cell division that then occur, give rise to the emergence of form, so that morphogenesis and differentiation become prime targets for external stimuli and chemical regulators.

But during their evolution, higher plants have elaborated growing regions or meristems to which the functions of cell multiplication and tissue differentiation have been primarily assigned (Wardlaw, 1965b). In these apical, lateral, or axillary growing regions organogenesis may be repetitive and vegetative growth is "open-ended" or even indefinitely indeterminate.

Preoccupied as one may be with a specific chemical reaction of a growth regulator at an intracellular site, its consequences are usually to be seen in the intercellular redistributions of the growth that ensues (Cutter, 1965; Wardlaw, 1968); this involves changes in an already tight system of localized regulatory controls which limit *in situ* the otherwise free capacity for growth and random proliferation of cells. These controls are those which, in normal development, ensure the emergence of form and division of labor in the plant body. Since we can explain so little about the causes of normal development and morphogenesis, the observations on exogenous growth regulatory substances have often been made empirically, but the disturbances

Fig. 2-9(A). For legend see facing page.

so created, contain clues to the ways growth is normally correlated and controlled.

Rest and dormancy, the formation of organs of perennation, the progressive development through juvenile and adult forms, or the transition from the vegetative to the flowering condition are all responses to morphogenetic stimuli mediated by, if not initially triggered by, chemical substances.

Growing Regions

The study of shoot apices in the development of vascular land plants has engaged fully some of the best minds in botanical science, as documented in such works as those of Sachs (1887), Schüepp (1966), Esau (1965), and

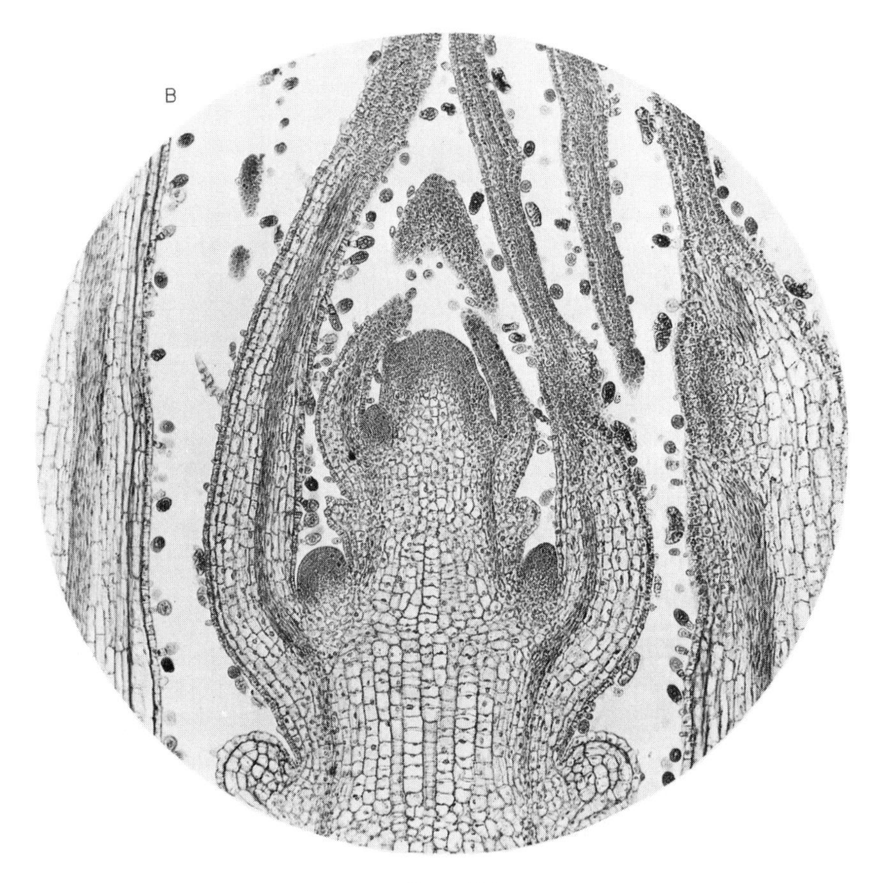

Fig. 2-9(B)

Fig. 2-9. The shoot apex of mint (*Mentha piperita*). A, transverse section showing the central apical dome and successive pairs of opposite, decussate leaves; B, longitudinal section showing leaf primordia, procambial strands, and axillary buds (Howe and Steward, 1962).

Clowes (1961), to mention but a few. Any observations here can, therefore, only point to a few salient ideas.

Wardlaw reproduced to the same scale photographs of the vegetative shoot apex of many primitive and advanced land plants. The range of size and form differs greatly in more primitive land plants. He pointed out that the entire shoot apex of a horsetail could fit into the large apical cell in the growing tip of a fern (Wardlaw, 1953). There are, however, evident differences in the form of the shoot apex and in the distribution of growth by cell division in primitive land plants and in angiosperms. In flowering plants, there is a more uniform pattern in their apices although, as they grow, great diversity of leaf form and arrangement develops.

The striking feature of plant meristems is the small size of the formative region. The more or less hemispherical central dome of cells in a shoot tip is often very small; in *Lupinus* it may only weigh about 0.0016 mg though it contains about 3500 cells (Sunderland and Brown, 1956). The central apical dome in a mint plant is shown in Fig. 2-9. Earlier it was thought that the surface of this central part of the apex was the main site of the cell divisions which produce the vegetative plant body, but this is now recognized not to be so (Clowes, 1961). Evidence from attempts to culture the central apical dome of angiosperm shoot tips, deprived of all primordia of lateral appendages and of the subjacent tissue in the axis, in an organized way have ordinarily failed, even when supplied externally with all known nutrients and stimuli (see Cutter, 1965; Nougarède, 1967). Nevertheless, a recent report (Smith and Murashige, 1970) makes the claim with tangible evidence that the meristematic dome tissue of several angiosperms with no visible leaf primordia can develop on a relatively simple basal medium supplemented with auxin and *myo*-inositol. This achievement, if substantiated, is all the more remarkable since the formative areas on the flanks are exceedingly difficult to exclude during the excision process. The lateral organs of the shoot really originate as a result of the localized activity of initiating cells on the flanks of the shoot tip and these produce the leaf primordia. Transverse divisions of subjacent tissue, i.e., the rib meristem, produce the central axis. For reasons yet unknown, dicotyledonous plants (though not monocotyledons) retain a series of strands, which form a ring of cells in the axis that continue to divide, i.e., the procambial strands, as well as isolated pockets of cells in the axils of leaves that become buds (Esau, 1965). Thus the processes of cell growth and cell division need to be locally turned on or off in meristems to explain the obvious facts of development. If a shoot segment is large enough to include leaf primordia and some subjacent tissue, there is usually no problem in securing its continued growth in culture and, especially so, if the segment should develop root tips. The virtually perpetual ability of clones of some roots (especially of dicotyledons) to grow, if their tips are repeatedly excised, is well established (Street 1957, 1969), although it is obvious that the small segments so excised must include the entire formative region of the root tip. It would be interesting in this context if the culture of root tips, with or without their "quiescent regions" of Clowes (1961), could be tested. If a small segment from a shoot apex is caused to proliferate and form a callus, it may often become a means of propagation as it forms adventitious buds. If, instead, a proliferating meristem segment is induced to form a free cell culture, the possibility that the free cells may develop into embryos and plantlets in large numbers may now be entertained (Steward *et al.*, 1970). In all of this, it is obvious that the growth of the cells *in situ* must be subject to localized regulatory control. The very different ways in which the shoots of dicotyledonous and monocotyledonous plants grow also present very different target sites for applied growth regulatory substances as evidenced by the ways in which some, e.g., 2,4-D, differentiate between broad and narrow leaved plants (Audus, 1964). Monocotyledons lack cambium in their vascular strands, have intercalary growing regions which are prominent at the base of leaves, and their encircling leaf bases form the main axis (Esau,

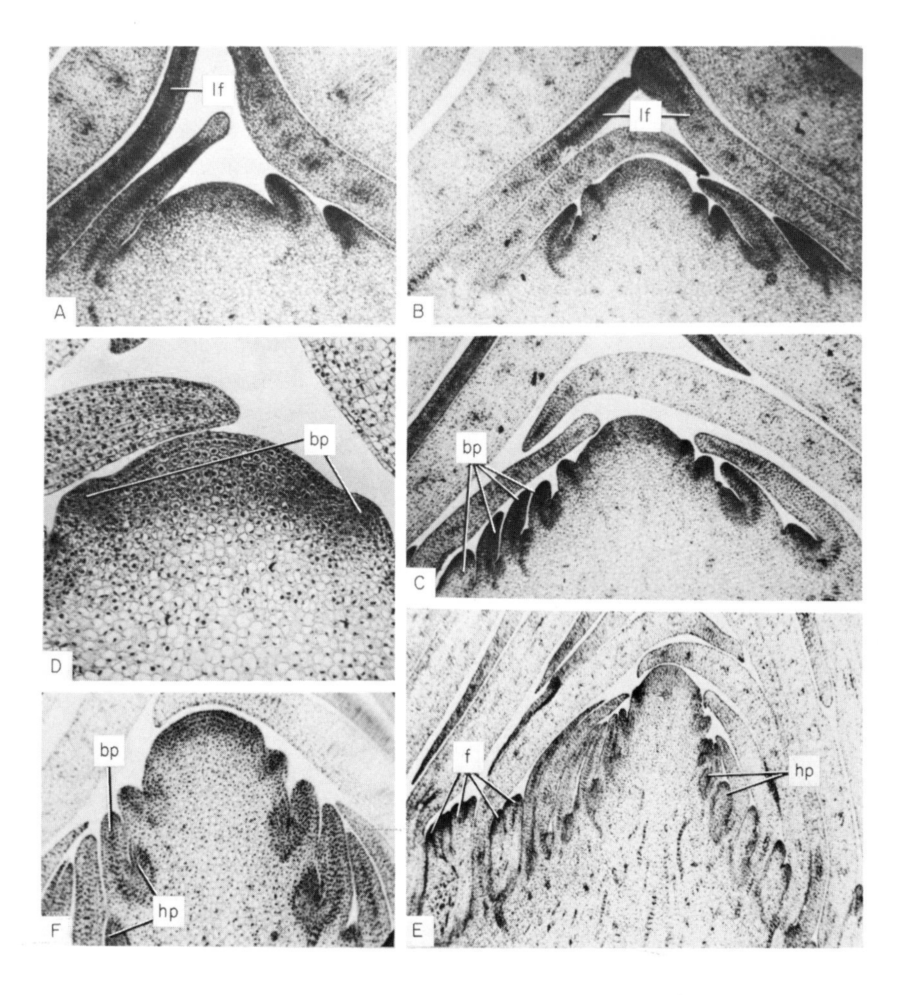

Fig. 2-10. Transformation of the growing point of the banana plant from the vegetative to the flowering state. A and B, vegetative apex with encircling leaf primordia (*lf*); C and D, presence of floral bract primordia (*bp*) in the axils of which flowers in "hands" (*hp*) appear in E and F. Flower primordia (*f*) are shown in E. Note the change in shape and elongated growth of the tip of the flowering shoot (Mohan Ram *et al.*, 1962).

1965). The rosette habit and basal meristematic growth of many monocotyledons explains their resistance to mowing, grazing, and many herbicidal substances.

Division rates in the central dome of shoot apices have been estimated at one division every third day, or even once every five days in *Lupinus*, but cell division seems to occur with much greater frequency in root tips (Richards, 1951). There seems little doubt, moreover, that small isolated tissue explants under the most favorable conditions can support cells that divide more frequently than the cells in many shoot apices (Steward, 1968).

As localized centers of growth appear in the shoot apex to become leaf primordia, they tend to grow, in a measure, independently of the main axis. They seem to arise far apart, as if to avoid competition with each other (as in the opposite and decussate leaves of mint in the apex shown in Fig. 2-9A). Each leaf, or whorl of leaves, obligates the activity of the growing tip at a given level for a given formative period until the next leaf, or whorl, repeats the process through the activity of similar initiating cells which arise elsewhere on the surface of the shoot tip. The time interval required for the apex to produce one leaf primordium and the internode that subtends it is a "plastochron" (Richards and Schwabe, 1969). Under given conditions of growth this interval for lupin was about 2 days (Sunderland *et al.*, 1956) and in mint was about 3.2 days (Howe and Steward, 1962); obviously, however, the plastochron will be a function also of the diurnally fluctuating conditions that modify the form of shoots. Moreover, the great contrasts between the development of shoots and roots (with the superficial lateral appendages of shoots contrasted with deep-seated ones in roots; and with the contrasting directions of primary xylem development and consequential arrangement of tissues in the axis) are apparent (Esau, 1965), as well as the dramatic changes that occur in the transition from vegetative to flowering shoot apices (Nougarède, 1967; Wardlaw, 1965b,c, 1968) (see Fig. 2-10). If one considers development to be the consequence of localized controls over the growth of cells, one realizes the formidable problems that are entailed in its causal interpretation. The innumerable ways that applied chemicals intervene to modify this normal pattern of behavior are, understandably, more easily observed than explained.

CHAPTER 3 *History and Modern Concepts of Growth-Regulating Substances*

Genesis of the Problem

For all practical purposes, the modern problem of the chemical regulation of growth of plants stems from Charles Darwin and his studies of the "Power of Movement in Plants" (1880). This statement is made despite suggestive and much earlier observations (Darwin, 1966). As early as 1754, Bonnet attempted to grow excised embryos of flowering plants, with and without portions which were removed, in crude solutions and, even earlier, Malpighi appreciated the nutritive role of the cotyledons in the development of embryos (Adelmann, 1966). However, the definitive conclusion of Darwin, based on his work with *Phalaris*, is contained in the phrase, "We must therefore conclude that when seedlings are freely exposed to a lateral light some influence is transmitted from the upper to the lower point, causing the latter to bend" (see Darwin, 1880, p. 474). From these origins, the work on the coleoptile of grasses proceeded and the essential steps in its development were ably summarized by Boysen Jensen (1936) in the diagrams of Fig. 3-1 reproduced here from the Avery and Burkholder translation of his work "Growth Hormones in Plants." The charts show how the evidence was mobilized to prove that the coleoptile tip produces a chemical substance which is transmitted basipetally and stimulates the elongating cells to grow. It was also shown that unilateral stimulation by light or gravity changes the distribution of this substance to produce the asymmetric growth which actually occurs. Boysen Jensen proved that the "influence" could move through a layer of gelatin, but that it could not

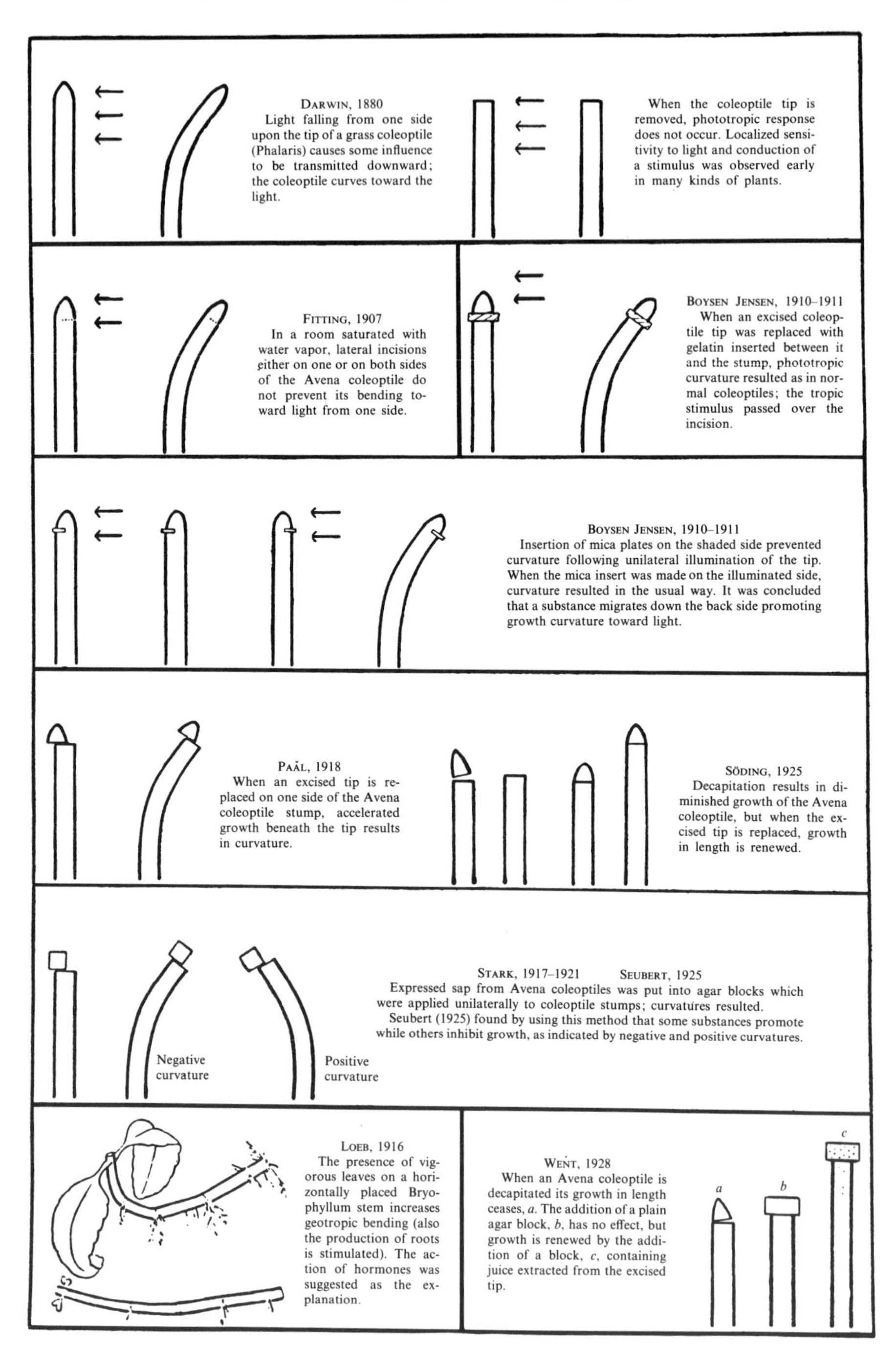
DARWIN, 1880
Light falling from one side upon the tip of a grass coleoptile (Phalaris) causes some influence to be transmitted downward; the coleoptile curves toward the light.
When the coleoptile tip is removed, phototropic response does not occur. Localized sensitivity to light and conduction of a stimulus was observed early in many kinds of plants.
FITTING, 1907
In a room saturated with water vapor, lateral incisions either on one or on both sides of the Avena coleoptile do not prevent its bending toward light from one side.
BOYSEN JENSEN, 1910–1911
When an excised coleoptile tip was replaced with gelatin inserted between it and the stump, phototropic curvature resulted as in normal coleoptiles; the tropic stimulus passed over the incision.
BOYSEN JENSEN, 1910–1911
Insertion of mica plates on the shaded side prevented curvature following unilateral illumination of the tip. When the mica insert was made on the illuminated side, curvature resulted in the usual way. It was concluded that a substance migrates down the back side promoting growth curvature toward light.
PAÁL, 1918
When an excised tip is replaced on one side of the Avena coleoptile stump, accelerated growth beneath the tip results in curvature.
SÖDING, 1925
Decapitation results in diminished growth of the Avena coleoptile, but when the excised tip is replaced, growth in length is renewed.
STARK, 1917–1921 SEUBERT, 1925
Expressed sap from Avena coleoptiles was put into agar blocks which were applied unilaterally to coleoptile stumps; curvatures resulted.
Seubert (1925) found by using this method that some substances promote while others inhibit growth, as indicated by negative and positive curvatures.
Negative curvature
Positive curvature
LOEB, 1916
The presence of vigorous leaves on a horizontally placed Bryophyllum stem increases geotropic bending (also the production of roots is stimulated). The action of hormones was suggested as the explanation.
WENT, 1928
When an Avena coleoptile is decapitated its growth in length ceases, *a*. The addition of a plain agar block, *b*, has no effect, but growth is renewed by the addition of a block, *c*, containing juice extracted from the excised tip.
a
b
c

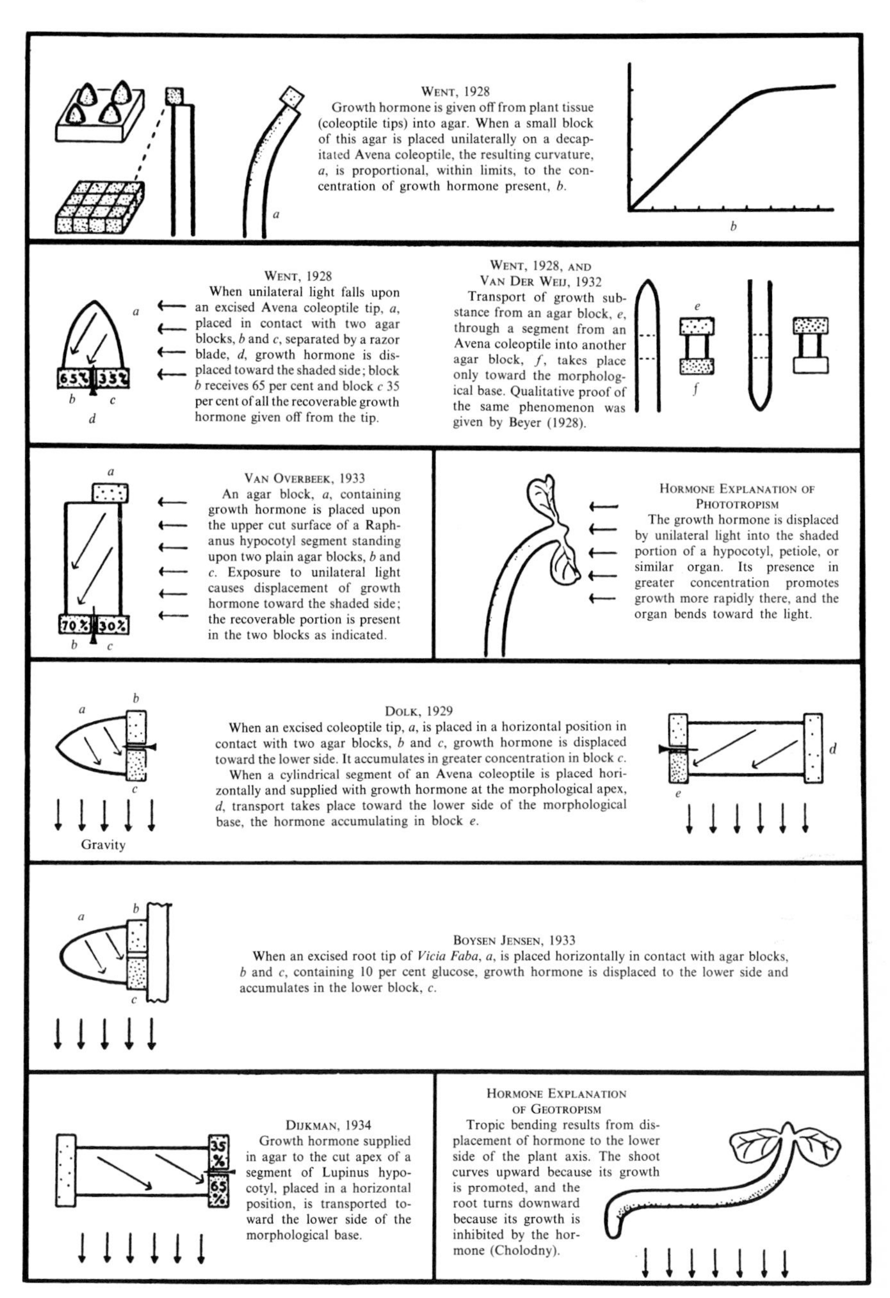

Fig. 3-1. Historical outline of the early discoveries concerning plant growth hormones. [From Boysen Jensen (1936).]

pass through material impervious to water such as mica. Páal in 1918 repeated and extended the experiments and showed that the growth of a particular organ, the hollow sheath or coleoptile within which grass seedlings emerge (see Fig. 2-6), is controlled from its tip through the agency of a diffusible substance. About 1928, Went showed that this active agent diffused into gelatin from living oat coleoptile tips; the agent which had diffused into the gelatin stimulated cell elongation just as it did when in living tips. By a series of elegant but simple experiments the evidence became conclusive, and investigators like Went and Thimann perfected the assays which enabled them to recognize and assess the activity of substances which had come to be called (following Kögl's suggestion) auxins (from the Greek, *auxein,* to increase). The initial extracts of auxin-like substances were made from urines (animal and human) but the definitive development came when Thimann isolated auxin from the tryptophan-containing medium in which the fungus *Rhizopus* was grown. The active substance, first named "rhizopin" by Nielsen, was shown to be identical with the indole-3-acetic acid (IAA) that had been isolated from normal and pathological urines. [For a detailed history of the above developments, reference may be made to Boysen Jensen (1936), Thimann (1948, 1952), F. A. F. C. Went (1935), F. W. Went (1935), and Went and Thimann (1937).] However, IAA was not isolated from flowering plant sources until 1946 when it was obtained from extracts of immature maize kernels, i.e., corn in the "milk" stage (Haagen-Smit *et al.,* 1946). Early in this period, however, IAA as *the* auxin, as *the* plant growth hormone (F. W. Went, 1935), became increasingly implicated in other regulated responses, such as root formation, flowering and fruit formation, abscission, cambial activity, seed germination, etc. (Thimann, 1948, 1952). The later use of chromatographic techniques has permitted the recognition of many substances that are now classified as auxins because of their activity in the auxin bioassays (Mitchell and Livingston, 1968; Morré and Key, 1967). The ensuing developments in the production of analogs of indole acetic acid received a great stimulus during World War II and, from this, stemmed the synthesis and use of the naphthoxyacetic and phenoxyacetic acids as auxin-like growth regulators (see Peterson, 1967).

The foundations of another major historical trend, now evident, although at first unnoticed in the West, were laid in Formosa in 1926 in a paper published by the Japanese plant physiologist Kurosawa on what are now known as the gibberellins (Kurosawa, 1926). This trend was to lead through concepts of vernalization, photoperiodism, and flower-forming substances to a currently broad area of chemically controlled plant development.

Kurosawa recognized that the toxic effects of the Bakanae fungus (associated with the "foolish seedling" disease) on rice were attributable to a secreted toxin, peculiar to the fungus (*Gibberella fujikuroi*, i.e., the sexual stage of *Fusarium moniliforme*). However, it was not until 1935 through the intensive work of a series of Japanese investigators that the isolation of a crystalline active material, named gibberellin, was announced by Yabuta (Stowe and Yamaki, 1957). These developments remained unnoticed outside of Japan because of the circumstances prior to and during World War II (Stodola, 1958). The technology, which was not available to the first Japanese workers for mass culture of the fungus and large-scale fractionations of extracts, became available after the war (as a by-product of wartime developments in the antibiotic and fermentation industries) to various groups. The activities of three main groups (in Japan, in Illinois, and in Imperial Chemical Industries in Britain) led to the isolation of different, but related, pure products, but it was not until 1954 that the structure of one of these, gibberellic acid (GA_3), was announced (Brian *et al.*, 1960; Stodola, 1958). As in the case of IAA, first isolated as a fungus metabolite, the evidence for gibberellin-like substances in flowering plants was long delayed (Phinney *et al.*, 1957). But now, a great array of these complex terpene-like substances are known as fungal metabolites on the one hand, and as physiologically active, regulatory substances of flowering plants on the other (Lang, 1970).

Another, even earlier, trail was to merge eventually with the knowledge of, and work on, the gibberellins. This trail may be traced to Sachs' early concepts of organ (flower)-forming substances which, however in chemical terms, could not then mature (see Sachs, 1887, p. 534). Chailakhyan points out (1968b) that Sachs lacked the knowledge of the specific conditions which were necessary to start flowering, an essential prerequisite to precise understanding of the chemical stimuli involved. This degree of precision, as it exists at present, only emerged from work and concepts developed over a long period of time. Sachs' basically correct suggestions of specific plant regulatory substances were ignored by subsequent workers, but he is essentially vindicated today. It was Klebs who appreciated the role of nutrients and environment in the balanced control of flowering, as in the concept of "ripeness to flower" (Klebs, 1918). Gassner in 1918 recognized the need of winter wheats for cold in contrast to their spring-sown counterparts (see Whyte, 1948). The consequences of chilling in the seedling stages of normally biennial crops for their later development, were appreciated by Thompson (1953, and references there cited) who, in effect, anticipated the doctrine of vernalization as it was later applied to cereal grains by Lysenko. The term vernalization, the English cognate for the Russian "jarovizacija" meaning "to make spring again," described treatments which convert normally summer-sown biennial wheats into a condition that allows them to flower in one season and to become suitable for sowing in spring (see Whyte, 1948, for a historical detailed account of these developments). Both the effects of abnormally low and maintained high temperature on plant development were attributed by Curtis to the growing points as the site of action (Curtis and Chang, 1930), but without any knowledge at that time of the chemical basis of perception and transmission of the cold stimulus. This phenomenon of growth regulation by temperature furnished the link with the regulatory role of gibberellins as substances that substitute for an otherwise necessary period of exposure to cold, as in the onset of flowering in biennial plants (Lang, 1957).

Another historical trend needs to be traced through the progressive understanding of flowering as it is affected by environmental factors. Although the doctrine of photoperiodism is rightly attributed to Garner and Allard (1920) who formulated it clearly, it was anticipated by Henfrey in 1852, by Liberty Hyde Bailey in 1893, and more precisely by Tournois just prior to World War I (see Evans, 1969b). The preoccupation with photoperiodism diverted attention from the role of temperature as seen in vernalization phenomena (see Murneek, 1948), and only recently has the full scope of the interplay of diurnally fluctuating cycles of light and temperature been revived. The first concept of the flower-forming substance of Sachs and the tacit recognition by Chailakhyan between 1933 and 1937 that flowering was as attributable to a flowering hormone (florigen) as the phenomena of growth were to auxins (Chailakhyan, 1937) found its experimental justification in grafting experiments that transferred the photoperiodically induced stimulus from one shoot to another (see Fig. 3-2A). This evidence necessitated some tangible chemical stimulus to affect the growing regions. The simple concept of a single flowering substance was modified however (Chailakhyan, 1961). Chailakhyan formulated views which attributed the flowering stimulus to the dual action of gibberellinlike substances necessary for stem elongation, and a class of anthesins necessary for flower formation. It is the balance between these component parts of the system that is now construed by Chailakhyan to be the florigen complex (1964, 1968a,b) (see Fig. 3-2B). But it should not be forgotten in all this that, as Garner and Allard appreciated, the environmental variables triggering off flowering also set in motion many other morphogenetic events that must equally have a metabolic basis and chemical causation. These events range from tuber and bulb formation to the onset of dormancy in axillary and terminal vegetative buds (Hendricks and Borthwick, 1963).

The conspicuous observations on light-mediated, morphogenetic responses (listed in Table 3-1) require a pigment as the means of perception of the stimulus. The development of the concept of light perception and the search for the necessary pigment (later called phytochrome), and the

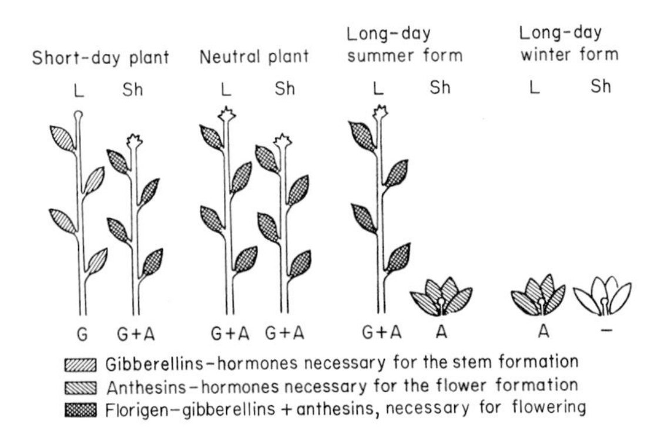

Fig. 3-2. The concept of the flowering stimulus after Chailakhyan. Top, flowering of *Rudbeckia* under short-day conditions. Left, *Nicotiana sylvestris* plant grown under long days to yield an extract applied to *Rudbeckia*; center, *Rudbeckia* plant stimulated to flower under short days by the *N. sylvestris* extract; right, control *Rudbeckia* plant untreated and grown under short days. Bottom, scheme of formation of flowering hormones in various plant species; L denotes long-day conditions; Sh, short-day conditions; G denotes gibberellin-producing; A, anthesin producing. [A, from photographs supplied by Dr. M. Kh. Chailakhyan, Timiriyazev Institute of Plant Physiology, Moscow; B, from Chailakhyan (1961).]

TABLE 3-1
SOME RESPONSES WHICH MAY BE MEDIATED BY PHYTOCHROME[a]

Flower initiation	Rhizome formation	Seed germination
Flower development	Bulbing	Fern spore germination
Formation of cleistogamous flowers (i.e., fertilized in the unopened bud)	Leaf enlargement	Anthocyanin formation, Gemmae production
Sex expression	Internode elongation	Epinasty (change in the inclination of petiole to the axis)
	Plumular hook unfolding	Dormancy in buds
	Chloroplast orientation and development	Leaf abscission

[a] After Hendricks and Borthwick, 1963.

understanding of its relations to the wavelength of incident light is a now familiar and distinguished chapter of modern plant physiology (Siegelman and Hendricks, 1964; Hendricks and Siegelman, 1967). Nevertheless, the means by which the stimulus so perceived (as in leaves) is transmitted and elicits a given response at the site of action is part of the "central mystery." This was aptly stated by Evans as follows: "The dogmas are that one pigment, phytochrome, mediates the initial photoperiodic reactions in all plants, and one hormone, florigen, concludes them. The central mystery is, with a common beginning and a common end, the intermediate reactions require darkness in short day plants and light in long day plants" (Evans, 1969b, p. 9).

Even as early as the late nineteenth century, other apparently unconnected inquiries were to anticipate and lead to many current developments in the chemical regulation of growth in plants. In 1892 von Wiesner published a little appreciated monograph concerning the basic structure and growth of the "stuff of life." Wiesner focused attention upon the phenomena of the transformation of resting cells to meristematic ones and of wound healing and clearly suggested that translocatable "chemical substances" stimulated the cell divisions (Wiesner, 1892, pp. 104–105).

Haberlandt (1913), sometime later, became involved in the investigation of wound healing by a rather circuitous route. He had previously attempted to maintain and culture surviving isolated cells of angiosperms (see Krikorian and Berquam, 1969). Although he had confidently prophesied that this would eventually be possible, his own efforts toward this end were frustrated due to the techniques of the day. Accordingly, Haberlandt turned to the use of tissue slices in which he knew that cells at the surface would divide in the formation, as in the case of potato tuber, of a regenerated periderm. He investigated this system in depth to ascertain what he could about the factors that induced the cells to divide and made the following interpretations of the phenomena he observed, although they were at that time unsupported by any direct chemical evidence. Cell division, in potato tuber, was stimulated, according to Haberlandt, by two substances; one, the substance lepto-hormone, derived from the vascular tissue (phloem), and the other, the wound hormone proper, was secreted by the injured cells (Haberlandt, 1921). It is appropriate to note here, however, that in 1909 Fitting had already found that a substance present in orchid pollen, later shown to be auxin, caused swelling of the gynostemium in the orchid flower. The word "hormone," borrowed from Starling, who used it in 1906 (Huxley, 1935), was used for the first time in the botanical literature in this connection in 1910 (Fitting, 1910, p. 265).

In retrospect, it is easy to see how attention was diverted from the more difficult problem of cell division, to the apparently more amenable one of cell enlargement, as this could be investigated through auxin-induced elongation of coleoptiles.

A brief, but abortive, interlude was associated with the investigations of English *et al.* (1939a,b) into a wound substance which they named traumatin or traumatic acid (1-decene-1,10-dicarboxylic acid). This substance was assigned the ability to stimulate cell divisions in the form of intumescences on the inner lining of bean pods (Bonner and English, 1938), which were used as an assay system, but the identity of that substance with the role of the hormone has been neither generally accepted nor confirmed (Bloch, 1941; Strong, 1958).

In the early days of callus tissue cultures, the addition of auxin was a favored device to induce more rapid growth (Gautheret, 1942). Nevertheless, the well-known example of the "habituated" carrot cultures of Gautheret (1955) was one in which the tissue in culture eventually became autotrophic for growth regulators and independent of anything previously attributable to auxin. In his philosophical observations on cell division, Haberlandt had speculated upon the role that pollen tube exudates might have on the induction of cell division in ovary walls or on cells in hanging drop cultures, and he even went so far as to suggest that it might be a good idea to try the effect of embryo sac fluids (Krikorian and Berquam, 1969). In embryo culture, first carried out for genetic reasons to cultivate embryos that would otherwise abort, the associates of Blakeslee made use of the liquid endosperm of coconut in the culture of *Datura* embryos (van Overbeek *et al.*, 1941) devices that were later applied by Ball to cultivate apices of *Lupinus* and *Tropaeolum* (Ball, 1946).

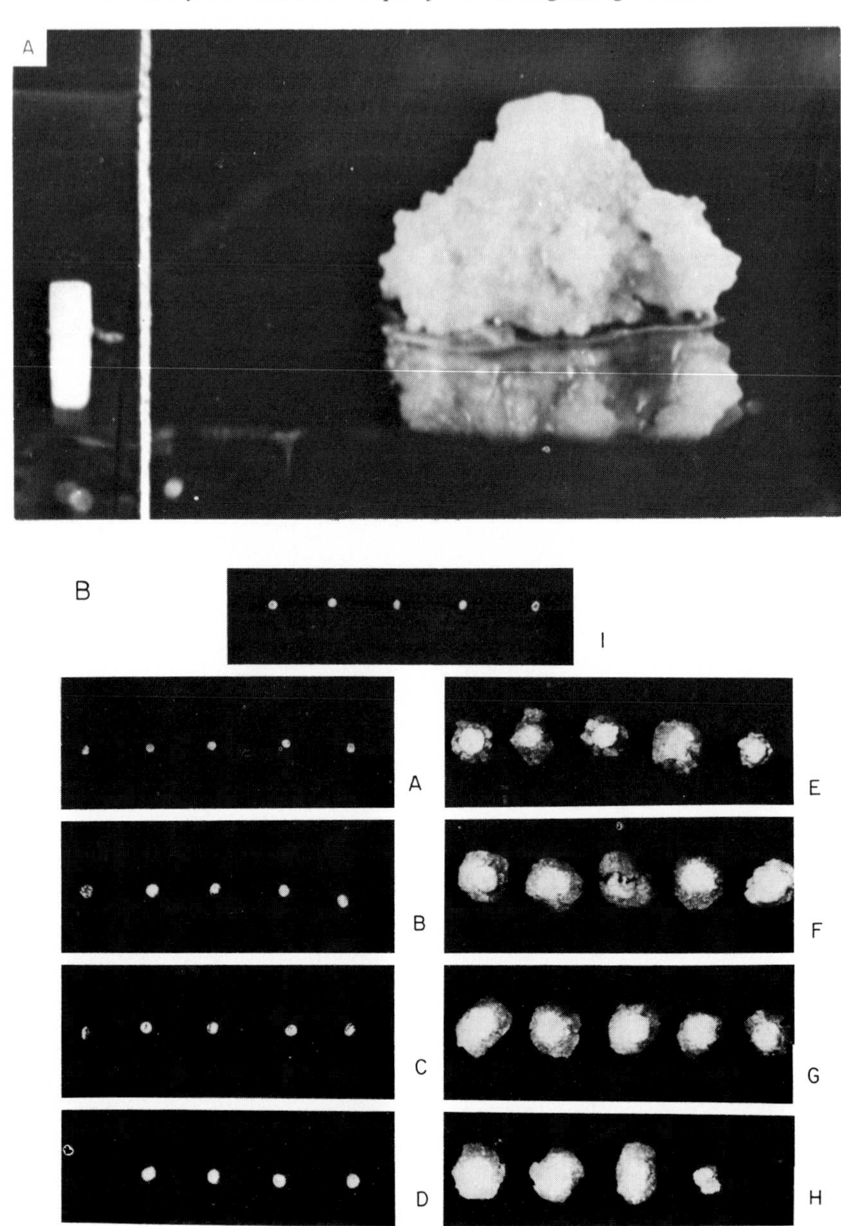

Fig. 3-3. A, contrasted effect of growth of a carrot explant on White's basal medium (left) and with coconut milk supplement (right). B, relative effects of indole-3-acetic acid (IAA) and coconut milk on growth of carrot explants; I, denotes initial size of explants; A–D, explants after growth on basal medium plus different concentrations (10.0, 1.0, 0.1, and 0.01 mg/liter, respectively) of IAA; E–H, same with 15% coconut milk added (Caplin and Steward, 1948).

The observation that coconut milk or coconut water added to an apparently complete tissue culture medium greatly stimulated growth and cell division of carrot phloem explants (Caplin and Steward, 1948) started a trend in the reinvestigation of substances and extracts that promote cell division (see Fig. 3-3). From this point, different kinds of assay systems and different classes of reputedly active substances have been extensively investigated from the standpoint of their respective roles in the stimulation or inhibition of cell division. Even the gas ethylene was credited long ago with the ability to induce intumescences, as on apple twigs (Wallace, 1926). In other words, the concept that over and above the role of nutrients, growth by cell division may be regulated by a balance of exogenous and mutually competitive substances seemed firmly established. Furthermore, the active substances in question were clearly homologized with the hormones of the animal body (see Thimann, 1948; Audus, 1963).

Problems of Terminology

The terminology of growth-regulating substances in plants may have leaned too heavily on the supposed analogies with animals, and this analogy in turn was heavily indebted to the phenomenon of "action at a distance" (Thimann, 1965). The concept of plant hormones, analogous to the secretion products of ductless glands in the animal body, was born despite many obvious dissimilarities in the way they act, or in the type of events that they control (Thimann, 1948). While these concepts are irretrievably embedded in the terminology and in the history of the subject, it is well to be aware that the similarities become misleading if they are pressed too far. This is obvious for reasons which are implicit in the very different organization and manner of growth and development of higher animals and plants, and in the very different nutritional organization of their constituent cells.

Plant growth regulatory substances, natural or synthetic, have traditionally been identified by certain operational terms; namely, as auxins (which by derivation cause an increase in size) (Thimann, 1969); as cytokinins which characteristically cause an increase in the number of cells (Skoog and Armstrong, 1970); as gibberellins, which were first prominent in effecting an increase in elongation growth of otherwise dwarfed plants (Lang, 1970); as the still hypothetical florigens which are supposed to mediate flowering (Chailakhyan, 1968b; Lang, 1965), as the so-called abscisins (Addicott and Lyon, 1969; Wareing and Ryback, 1970); and more recently, as the newly erected group known as morphactins (Schneider, 1970; Ziegler, 1970). A dormin should cause previously growing cells or organs to enter a period of quiescence, rest, or dormancy; an abscisin, should activate cells in the region known as the abscission layer (see page 59) which is involved in leaf fall, fruit drop, etc.; a morphactin should be a morphologically active substance. These various terms for classes of growth regulators do not describe chemical substances per se, for each term may apply collectively to substances with a baffling array of structures. Similar ends may often be achieved by very different chemical means, so that a given substance may, under different circumstances, cause responses which might place it in different operational classes. A given substance may behave in ways that classify it as an auxin in one situation, or as a cytokinin in another. The responses in question are rarely as "clean-cut" as the definitions require, for the substances often act as parts of a balanced network of interacting effects, which modulate the behavior of cells by acting upon them in a variety of ways.

An autotrophic, unicellular, green alga presents the minimum of specialization, since this must reside only in its internal compartmentation into organelles. Its growth could be said to be regulated by its nutrition, i.e., by the major inorganic nutrient elements and by the "trace elements," such as manganese, copper, zinc, molybdenum, boron, and chloride which are each needed in a very small amount, as well as by the exogenous sources of carbon (whether these are in the form of carbon dioxide or of other substances), and by such obvious characteristics of its environment, as the temperature, the light, and the degree of aeration which obtains. Thus, a free-living, green, unicellular alga should be involved to a minimum extent with exogenous growth-regulating substances. But, even such cells still present the problem that their biochemical processes, genetically rendered feasible, are so precisely programmed and coordinated within the compartmented cells that they proceed in an incredibly orderly way. This problem is now being interpreted in terms of the spatial arrangement of enzyme systems on the surfaces of such membranes as those of mitochondria and plastids, etc. (Korn, 1969; Pardee, 1968; Rothfield and Finkelstein, 1968). The trace elements, through their frequently demonstrable role in enzyme systems, may regulate growth whenever they are rendered externally limiting; but they are not usually regarded as growth-regulating substances in the context of this work. Whenever the endogenous formation of vitamins, which often act as coenzymes, is inadequate so that they must be supplied exogenously, they also may control growth as well as metabolism; but these substances also are not mainly in question here.

Although plants as a whole may be autotrophic, few cells grow autotrophically *in situ.* In meristems, the highly heterotrophic nutrition of cells is obvious and appropriate regulation must determine the pace, or the direction, of the genetically feasible reactions of developmental processes which occur. When the "message" is communicated from one organ to another, as in the auxins, the hormone analogy of "action at a distance" is relevant; but where the "message" is communicated by one or several substances in an ambient culture medium for free cells it may be less obvious.

Many now seek explanations for all growth regulatory actions in terms of a means of "turning genes on or off" (Britten and Davidson, 1969). The unifying principle here is that causal substances affect nuclear genes and determine the level of their activity. Others, unwilling to go so far, will see exogenous substances, and indeed environmental stimuli, as able to intervene at many possible sites. Such stimuli may even interact with various autonomous organelles, without the implicit need to "ask the permission of the nucleus" for each and every act stimulated (Bogorad, 1967). This is the dilemma in all problems of morphogenesis and development where there is need to visualize epigenetic controls over the behavior of organized totipotent cells.

Auxins

F. A. F. C. and F. W. Went seem to have made the choice to first investigate the exogenous substances that promote growth by cell enlargement or, as in the case of the coleoptile of grasses, by growth in length. Out of their inquiries came the oft-quoted aphorism "no growth without growth substance" (Went, 1935). Despite their long history (see pp. 27–30), and the many observations of responses due to indole-3-acetic acid and other natural auxins (Bentley, 1958; Galston and Purves, 1960; Thimann, 1948, 1952, 1969) (see Fig. 3-4), there is no clear explanation of how and where the auxins act in cells. So much so that a recent and almost frenetically active area of research has developed around the observations of Burg that implicate the very simple molecule, ethylene, in either the causes or the consequences of auxin (IAA) action (Burg and Burg, 1968; Pratt and Goeschl, 1969).

Inevitably, a given assay system (see Mitchell and Livingston, 1968) detects physiologically active substances which have particular propensities, notably that they may be assayed by a given test. The characteristics of a class of compounds as assayed may, therefore, be more a property of the assay system used than of the full range of biological activities of the substances in question.

Fig. 3-4. Epinasty and copious root initiation caused by auxins on tomato plants. Top, left: untreated plant; rt: plant treated with lanolin paste plus 1% naphthaleneacetic acid. Bottom, detail of tomato shoot treated with lanolin paste containing 2% indole-3-acetic acid showing many adventitious roots. [Photographs courtesy of the Boyce Thompson Institute for Plant Research, Inc. (Zimmerman and Wilcoxon, 1935).]

Cytokinins

Similarly, one can identify substances as cytokinins by the use of a test system (Helgeson *et al.*, 1969; Letham, 1967b; Miller, 1963, 1967; Xhaufflaire and Gaspar, 1968) in which all other factors are relatively nonlimiting to growth, except some that are required to induce cell division (cf. pp. 50–51). But it may be an illusion that the substances which act exogenously as auxins in one assay, or others which act as cytokinins in another, do so endogenously solely in their own right; for, in assay systems that involve aspects of growth (whether by cell enlargement or by cell multiplication), many events at many sites must be set in motion simultaneously. Therefore, the more general and versatile the assay system becomes, the more it is apparent that one type of response, e.g., cell enlargement, may also lead to another (cell division) and vice versa.

Gibberellins

Conventionally these substances comprise a distinct class (Brian, 1966), although in many respects their role resembles that of auxins (Lang, 1966; Paleg, 1965). As already noted (p. 30–31) the first of these substances to be identified was gibberellic acid. The gibberellins are now prominently identified with their ability to cause elongation of genetically dwarfed plants (Phinney, 1956) and it is interesting to note that, these substances will convert Mendel's dwarf varieties of peas to tall (Brian and Hemming, 1955). A prominent, alternative role of these substances in development is their ability to replace the cold stimulus during the rest period of biennial flowering plants. Low temperature treatments cause some biennial plants to "bolt" and thus to flower prematurely (see Fig. 3-5); gibberellins will substitute for low temperature in this response. The gibberellins interact also with photoperiod in the determination of flowering (see Fig. 3-6). The effect of gibberellins is commonly regarded as a formative one; but, gibberellins may also act on previously initiated structures, causing them to expand and emerge (Lang, 1965, 1966) (see Fig. 3-7). Not unexpectedly, a large number of synthetic compounds are now known which are held to act upon plants by suppressing gibberellin production (Lang, 1970).

Hence, the present point of view is as follows. In all of the assays for plant growth regulators which are based upon the responses of preformed organs, the immediate stimulus merely "unblocks" some previous limita-

Fig. 3-5. The effect of exposure to cold in seedling celery plants (on the left) contrasted with those (on the right) not so exposed. Both sets were subsequently grown at the same temperature. The cold-treated seedlings "bolted," i.e., they formed flowering shoots. (Photograph courtesy of Professor H. C. Thompson, Cornell University.)

Fig. 3-6. The role of gibberellic acid in flowering: substitution of the long day stimulus in *Rudbeckia*. Left, plant treated with GA_3 and grown under short days; right, control plant grown under short days (see Fig. 3-2). (Photograph supplied by Department of Floriculture, Cornell University.)

Fig. 3-7. The replacement of a cold stimulus in *Rhododendron* by gibberellin. Left, unchilled flower bud which would require 8 weeks at 40°–50°F to open; center, comparable bud treated directly with 1 drop 0.1% gibberellic acid; right, comparable bud which received 1% gibberellic acid via the vascular supply beneath the bud. (Photograph courtesy of the U. S. Department of Agriculture.)

tion, but, in action, the molecules in question function as a part of a matrix of interacting, interlocking events. This philosophy is very different from the classical concept of hormonal action in animals. Growth regulators may now be considered in the context of what we know is needed to induce growth in otherwise quiescent tissue. Having considered this, having in effect unleashed the "built-in capacity for growth," one should trace the requirements for exogenous substances that regulate that growth, develop its maximum rate, and control its directions in morphogenesis.

CHAPTER 4 The Induction of Growth in Quiescent Cells

The ease with which cells pass from the most actively growing state to a comparatively mature quiescent state is an obvious feature of their development. Whether this transition is viewed spatially in terms of cell diameters, or in time, it is equally dramatic. Nevertheless, the mature parenchyma cells of pith or cortex, of tubers or bulbs, may have great longevity which, in extreme cases, e.g., certain cacti and woody trees, has been estimated in terms of hundreds of years (MacDougal, 1926; MacDougal and Long, 1927; MacDougal and Smith, 1927; Popham, 1958). This survival, and the storage functions which such cells may also discharge, implies a much reduced metabolic rate, which is best indicated by the low respiration of intact mature organs, even at temperatures normally conducive to much higher respiratory rates (see Table 4-1).

In part, these low metabolic rates may be seen as the consequence of form (e.g., the low surface to volume ratio of massive storage organs which limits exchange with their environment), or as the consequence of regulation by inhibitory substances which accumulate in senescent organs or quiescent cells (Steward and Mohan Ram, 1961). Concepts of inhibitors which accompany the onset of dormancy and disappear when it is broken are not new (Addicott and Lyon, 1969; Evenari, 1949, 1961; Giertych, 1964; Moreland *et al.*, 1966; Toole *et al.*, 1956), but they are now supported by the isolation and identification of substances which may play this role in buds, e.g., dormins (Addicott and Lyon, 1969); in tubers, e.g., solanidine (Kuhn and Löw, 1954); in bulbs, e.g., lycoricidinol (Okamoto and Torii, 1968); or in seeds, e.g., coumarin (Berrie *et al.*, 1968; Evanari, 1949; Toole *et al.*, 1956) and daphnetin (Moreland *et al.*, 1966; Toole *et al.*, 1956).

TABLE 4-1
THE PACE OF RESPIRATION

Organism	Temp. (°C)	$mm^3\ O_2$/gm wet wt/hour	$mm^3\ O_2$/gm dry wt/hour
Azotobacter spp.	28	—	200,000–400,000
Chlorella spp.	25	—	2,200
Euglena	25	800–1000	—
Baker's yeast	28	—	40,000– 80,000
Neurospora			
Dormant spores	25	90	250
Germinating spores	25	7,056	19,600
Carrot root	25	25– 30	—
Carrot leaves			
Young	22	1,133	—
Mature	22	439	—
Barley grains			
Dormant	25	0.06	—
Germinating	25	108	—
Mouse			
Resting	37	2,500	—
Running	37	20,000	—
Man			
Resting	37	200	—
Working	37	4,000	—
Elephant resting	37	148	—

[a] Rates are commonly expressed as $mm^3\ O_2$/gm fresh or dry weight. Some representative data are given. Note that respiratory intensity follows biological activity (compare barley grains dormant, with barley grains germinating). The respiration (or metabolic rate) of similar massive organisms or organs tends to reflect their proportion of surface to mass. Thus smaller organisms have higher respiratory rates per unit mass (compare mouse, man, and elephant). [After Goddard (1945).]

The Onset of Growth in Mature Cells

The events of "wound healing" at cut surfaces obviously involve the renewed proximity of surface cells to air and to moisture; the consequences upon metabolism of enhanced oxygen uptake; the conversion of storage substances, like starch, to soluble metabolites; the renewed synthesis of protein; the swelling and internal division of cells (Steward *et al.*, 1932); and all these in ways which may still entail the role of inhibitors in the quiescent state and invoke specific activitors of the induced state (see Fig. 4-1). It is difficult to distinguish causes from effects in all this (Bloch, 1941, 1952; Lipetz, 1970).

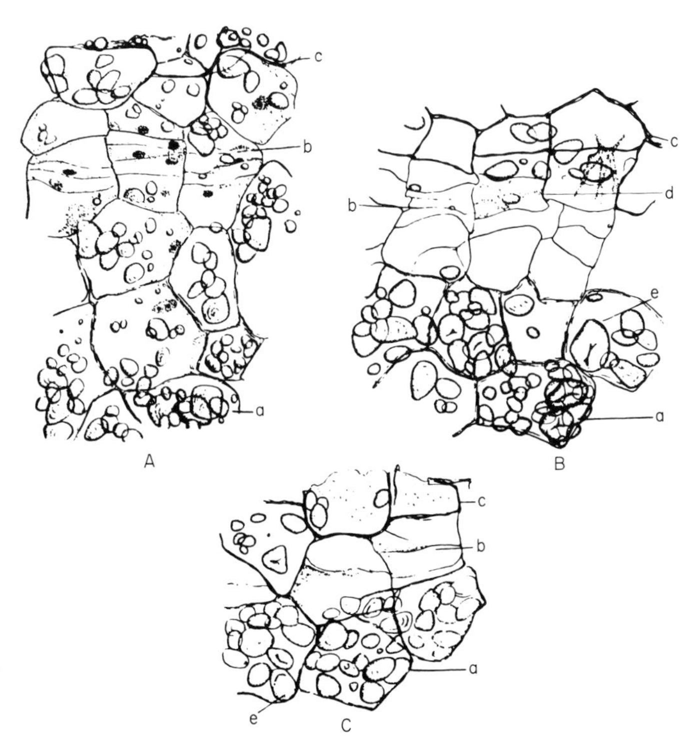

Fig. 4-1. The disappearance of starch and the occurrence of cell divisions at a cut potato tuber surface exposed to moist air. *a*, Parenchyma cells with an almost normal complement of starch grains; *b*, cork cambium (phellogen) formed by internal divisions within parenchyma; *c*, surface browning and suberization; *d*, parenchyma cell with prominent cytoplasmic strands along with vigorous protoplasmic streaming; *e*, cracking starch grains; A, discs in air; B, washed discs in air; C, discs in water (Steward *et al.*, 1932).

The range of responses displayed by different storage organs, from those which heal their wounds easily (the potato tuber) to those which do so with difficulty (pome fruits) or not at all (bulbs of monocotyledons), implies some profound differences in the endogenous ability of different preformed living cells to divide again.

A measure of the depth of the inhibitory control of the further growth of mature cells is the intensity of the stimuli needed to restore them to active growth, even under optimal conditions. Where these conditions first disturb the symmetry of intact organs, which had imposed prior restrictions on the volumes of cells, they first allow the cells to readjust their volume and they often do this rapidly as they respond to, as indeed they also supply evidence for, the role of auxins in the promotion of cell enlargement (Galston and Purves, 1960). Observations of this sort have ranged from those made

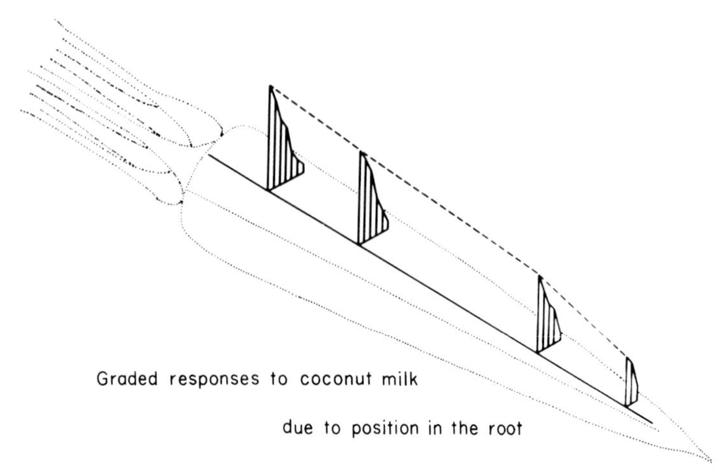

Fig. 4-2. Growth induction in carrot explants. Gradients in the responses to factors that induce growth. Standard carrot phloem explants were tested under standard assay conditions (see Fig. 4-8); the growth observed was a function of the original position of the explants along the axis of the root and also of radial distance from the cambium.

on the extending region of the coleoptiles of grasses to the curvatures of severed segments of seedling stems or hypocotyls, and they have produced a vast amount of knowledge on the structure/function relationships between the molecules which can intervene externally in this way (Bentley, 1961; Porter and Thimann, 1965; Veldstra, 1953; Wain and Fawcett, 1969). But where the stimuli to restore growth (as in pieces excised from carrot root) can lead to protracted and free cell proliferation, they have ranged from the casual use of auxins as an initial stimulus in the medium to the much greater growth that can now be incited by the controlled use of much more powerful stimuli, both natural and synthetic (Steward, 1968). Moreover, there are gradients in the plant body which determine the ease with which preformed cells of mature tissues and organs will respond in culture to the stimuli in question (Caplin and Steward, 1952), (see Fig. 4-2). Cells become chemically restricted in development in what they can do; this is also reflected by the nature and intensity of the stimuli needed to restore them to full activity.

The most chemically satisfying approach may seem to require the use of the assay system which responds in the most restricted and specific manner to the most limited range of chemical compounds. One can learn much in this way about the properties of specific chemical substances *in terms of a particular and restricted assay system.* The specificity of individual substances and the selectivity of different assay systems is apparent from such data as the following. In the *Avena* curvature test, indole-3-acetic acid (IAA) is approximately 1000 times as effective as 2,4-dichlorophenoxyacetic acid (2,4-D); on the other hand, 2,4-D is 12 times as effective as IAA in the split pea test, but in straight growth tests 2,4-D and IAA have comparable activities (see Wain and Fawcett, 1969, and references there cited).

But to interpret growth and the factors that regulate it, one also needs to investigate the responses of potentially totipotent cells and to observe them at very different levels of response. These cells are then affected by a great range of regulatory substances. Although the results obtained by these means may seem more frustrating, their eventual synthesis should be closer to biological reality.

Bioassays

The advantages and limitations of different bioassays for growth-regulating substances should be appreciated. These procedures for testing responses should be distinguished clearly from the now very well-developed procedures (e.g., thin-layer, gas, and other refined chromatographic and analytical techniques) which aid in the recognition and identification of the natural substances (Brinks *et al.*, 1969; Grunwald *et al.*, 1968; Hecht *et al.*, 1969c; Nishikawa *et al.*, 1966; Gaskin and MacMillan, 1968; Shannon and Letham, 1966). To appreciate the range of test systems that have been used, some are summarized in Table 4-2; the principal ones for auxins, gibberellins, cytokinins, abscisins, etc., are described in more operational detail by means of Figs 4-3 to 4-9 and their interpretive legends.

Whereas many of the bioassays utilize responses induced in preformed organs (see Fig. 2-6) or plant parts (see Fig. 4-10), the advantages of aseptically cultured tissues, in which growth may be both induced and maintained, for the bioassay of chemical growth factors is apparent.

Assays which are based on tissue culture systems, especially when they may give rise to totipotent free cells, present to growth regulatory substances systems which cover the full range of phenomena that growth and morphogenesis involve. This approach contrasts sharply with assays that are so restricted in their scope that they involve not only one kind of response (e.g., cell enlargement, etc.), but also one facet of the physiological role of the tested substances. For example, the coleoptile curvature assay (see Fig. 4-3) can only lead to the understanding of the role of IAA, or other auxins, in relation to cell enlargement (Thimann, 1969); it cannot possibly reveal the equally important role of these substances when they act in balanced systems that produce cell division. Similarly, the barley aleurone system (see Fig. 4-5) has revealed much about the role of gibberellins in relation to the induction of synthesis of α-amylase, or other enzymes (Filner

TABLE 4-2
SOME TYPES OF BIOASSAYS[a]

Plant material used	Test system	Parameter observed
	AUXINS	
Oat, coleoptile	Agar blocks containing test material placed asymmetrically on decapitated oat coleoptiles	Curvature
Oat or wheat, coleoptile sections	Uniform segments of coleoptiles	Length increments
Oat, first internode	Uniform segments of first internodes of etiolated oats	Length increments
Pea, stem sections	Uniform segments from third internode of etiolated peas (*Pisum sativum*)	Length measurements and weight change
Pea, split stem	Split stems of peas (var. *Alaska*)	Angle of curvature
Tomato or soybean, hypocotyl	Uniform segments of tomato hypocotyl	Length increments
Cress, root	Seedlings grown in various test concentrations	Growth rate (growth inhibition curve constructed)
Jerusalem artichoke, tuber slices or chicory root discs	Segments incubated on filter paper	Increase in cell volume as measured by fresh weight; mitotic frequency determined by microscopy
	GIBBERELLINS	
Dwarf pea, internode	Isolated or intact epicotyls	Elongation
Dwarf maize, leaf sheath	First unfolding leaves treated	Length of first and second leaf sheaths
Dwarf maize or dwarf bean, plants	Genetic dwarfs treated	Extension of the dwarfed stem
Cucumber or lettuce, hypocotyl	Young seedlings	Length of hypocotyl
Lentil seedlings, epicotyl	Excised shoots (of *Lens culinaris*)	Length of epicotyl
Barley, endosperm	Excised barley endosperm treated with test compounds	Release of reducing sugars or increase in α-amylase activity
	Excised barley aleurone layers	Increase in α-amylase or protease activity
Young wheat coleoptiles (insensitive to auxins)	Very young, whole, excised coleoptiles	Increase in length
Honesty, stem growth	Seedlings of *Lunaria annua*	Increase in stem length
	CYTOKININS	
Carrot, root phloem	Secondary phloem explants of *Daucus carota* subjected to test media	Increase in cell number; cell size; fresh weight; dry weight, etc.

TABLE 4-2 (*continued*)
SOME TYPES OF BIOASSAYS[a]

Plant material used	Test system	Parameter observed
	CYTOKININS	
Tobacco, callus or stem	Subcultured callus or pith explants	Fresh weight; dry weight; etc.
Soybean, callus or cotyledon	Subcultured callus or cotyledon explants	Fresh weight; dry weight; etc.
Barley, oat, tobacco, or other leaf material	Excised leaf sections or discs	Chlorophyll retention (loss prevented by cytokinins)
Mosses, e.g., *Funaria* or *Tortella*	Moss protonema or "leafy" gametophytes	Number of buds formed
Peas, mungbean, or other plants	Lateral buds of the treated plants	Increase in length of lateral buds
Radish and bean leaf	Leaf discs	Fresh weight and increase of disc area
Lettuce seeds	"Grand Rapids" variety germinated in darkness	Percentage germination in absence of red light
	ABSCISINS[b]	
Cotton, *Coleus* and bean, nodal explants	Explanted nodes are treated	Abscission of petiole
Birch, sycamore, maple, black currant	Youngest expanded leaf of seedlings and the apical region treated under long-day conditions	Inhibition of extension growth and formation of terminal buds
	MORPHACTINS[c]	
Wide range of plant materials too diverse to be specified	Seedlings oriented in predetermined or horizontal planes and exposed to a unilateral stimulus	Responses to gravity and light (loss of reactivity to gravity or a unilateral light source)
	Coleoptile tips placed horizontally into petri dishes	Degrees of curvature measured
	Shoots of pea seedling, or other plants, treated with a foliar spray	Inhibition of apical buds and stem elongation; stimulation of axillary buds (apical dominance weakened or abolished)

[a] Cf. Addicott *et al.*, 1964; Audus, 1963; El-Antably *et al.*, 1967; Helgeson *et al.*, 1969; Johri and Varner, 1967; Khan, 1967; Miller, 1963, 1967; Mitchell and Livingston, 1968; Morré and Key, 1967; Neely and Phinney, 1957; Went and Thimann, 1937.

[b] Many of the above tests may be, and have been, used as abscisin bioassays (lettuce germination, cucumber hypocotyl, and radicle extension, α-amylase inhibition).

[c] Like abscisins, morphactins intervene in many responses and may therefore be tested in a wide array of other assay systems.

et al., 1969; Filner and Varner, 1967; Varner and Johri, 1968), but it cannot equally contribute to the role of gibberellins in relation to flowering or to cell division which that assay does not involve. The cotton abscission assay (Addicott *et al.*, 1964, and Fig. 4-9) was used in ways which led to the isolation of the compound now known as abscisic acid (Addicott and Lyon, 1969), but, in fact, that compound was also isolated from sycamore (*Acer*) buds where its role as a dormin was most in question (Wareing and Ryback, 1970). The alternative names of abscisin II and dormin for what was the

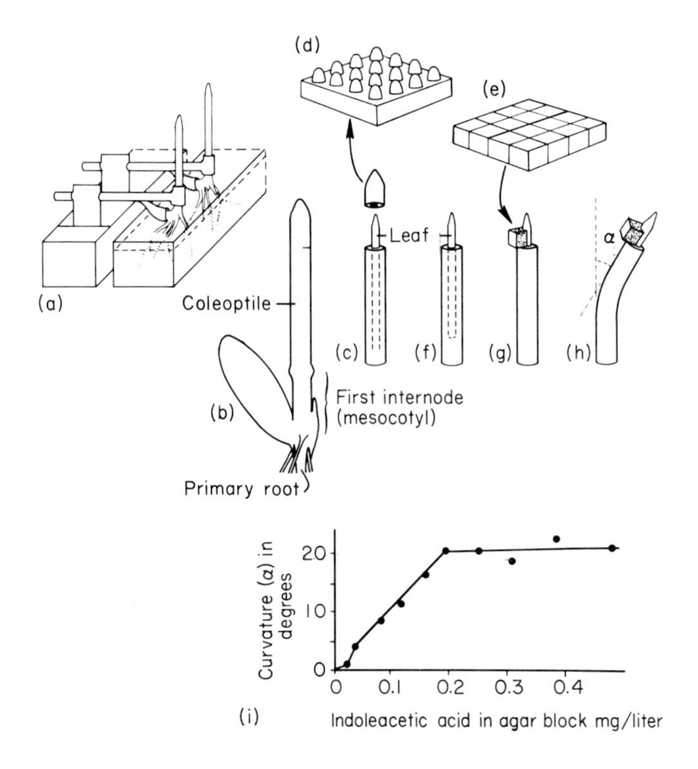

Fig. 4-3. Schematic representation of the *Avena* curvature test. Oat grains are germinated and grown in total darkness with specified short periods of red light. When the roots are about 2 mm long, the seedlings are "planted" in glass holders, using the water culture method (a). Straight seedlings (b) are selected and their tips are removed (c). Agar blocks containing the test substance (or for experiments of the diffusion type, the plant parts are placed directly on the agar) (d), and the agar is cut into blocks of standard size (usually 1 mm^3) (e). After "physiological regeneration of the coleoptile tip" occurs (about 3 hours), an additional 4 mm is decapitated (f), and the first leaf is pulled out to act as a support. The agar blocks are then placed on the decapitated coleoptile adjacent to the leaf (g). Shadowgraphs are taken after an appropriate period of time (90–110 minutes), and the angle of curvature (α) is determined (h). The relationship between curvature and auxin concentration is linear over a certain range (i). [Adapted from Went and Thimann (1937).]

same physiologically active substance, and the later compromise in the name abscisic acid (Addicott *et al.*, 1968) indicates that one should not too rigidly prescribe the physiological activity of a given growth regulator by a given assay used to demonstrate its activity. In the case in question (abscisic acid), it is now known that another isolate, from another plant (*Lupinus*), in the hands of other investigators (Cornforth *et al.*, 1966) who used another assay also led to the same substance.

Without question there has been a use for the highly restricted and specific assay systems which allows particular substances to be recognized. However, reliance upon these highly restricted systems so limits perception that the full range of possibilities may pass unnoticed. When the objective of a test system is to understand the multiplicity of factors that a growing system involves, there is merit if it can be responsive to the full range of factors that the genetic information allows. In this way, information for the behavior of the tissue or cells in isolation may have relevance to the behavior of the cells *in situ*. As experience with the carrot assay system has advanced, the study of many clones has given a wider appreciation of the

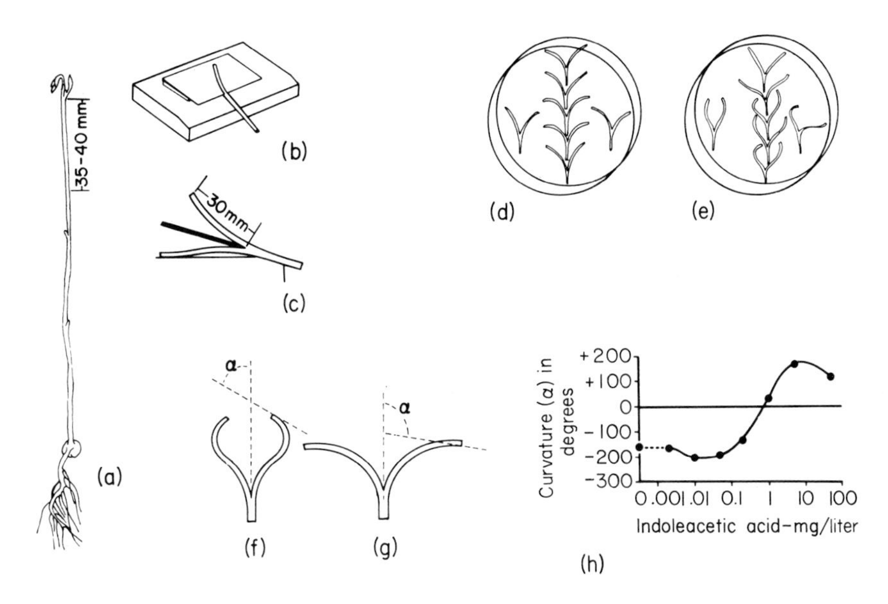

Fig. 4-4. Schematic representation of the split pea stem bioassay for auxins. Pea seeds germinated in complete darkness except for a daily 4 hour exposure to red light are harvested after 7 or 8 days (a). Plants with a third internode about 40–70 mm long are used for the test. Segments (35–40 mm) are excised and split with a special cutter (b) downwards about 30 mm (c). Split stems are then placed in petri dishes containing water (d) and solutions of growth regulator (e). Those in water remain unchanged but those in active substances bend inward or outward. Shadowgraphs may be made about 6 hours in the test solution. The angle (α) of positive curvature (f) and negative curvature (g) may then be measured. The relationship between curvature and auxin concentration (h). [Adapted from Thimann and Schneider (1938).]

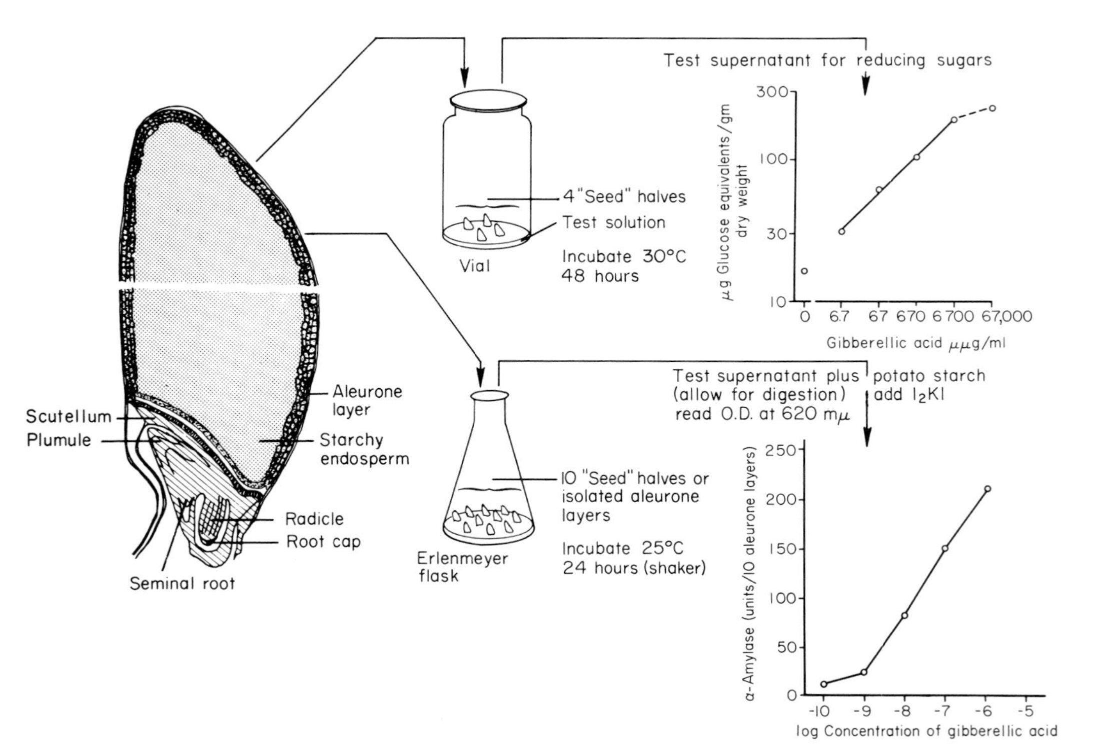

Fig. 4-5. The barley endosperm bioassay for determining gibberellin activity. Grains of any naked variety of barley (*Hordeum vulgare*) are sterilized and cut transversely as shown; the half containing the embryo is discarded; the halves are weighed and placed in vials containing test solution and incubated (ca. 48 hours). Reducing sugars released into the test solution may then be determined by any appropriate method. One alternative bioassay is performed with aseptically isolated aleurone layers or half-grains. These are again incubated in the test solution which is in turn tested for any increase in α-amylase activity by its effect on the digestion of starch. The release of enzymes such as proteases have likewise been shown to be affected by gibberellins (especially GA_3), and this is another parameter for a bioassay. [Adapted from Nicholls and Paleg (1963); Coombe *et al.*, (1967); Jones and Varner (1967).]

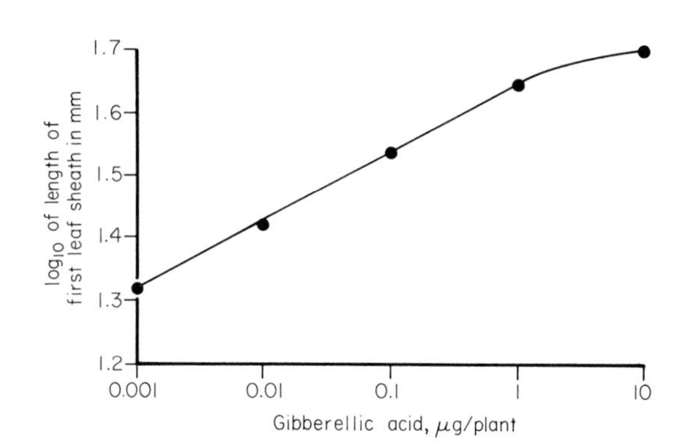

Fig. 4-6. The dwarf corn bioassay for gibberellins. There are at least six single gene mutants of *Zea mays* L. (dwarf-1 is shown here) which are valuable in studying the response to gibberellins. One technique involves applying a measured amount of test solution, in an appropriate wetting agent, into the first unfolding leaf of corn seedlings about 6 to 7 days old. The seedlings are then allowed to grow for another 6 to 7 days until the first and second leaves have fully matured. The length of the first leaf sheath is measured and the log of the length is plotted against the log of the dosage applied (after Neely and Phinney, 1957).

full range of properties that the carrot genome can encompass (Degani and Steward, 1968; Steward and Degani, 1969). By contrast, the tobacco assay system can be, and apparently has been, so progressively refined that it becomes so dependent on adenyl compounds alone (in the presence of an auxin) that it may seem these are the only compounds able to function as cytokinins, and, hence, the term cytokinin and adenyl derivative have become synonymous (Skoog *et al.*, 1967). This is clearly unacceptable and is the consequence of a too rigid reliance upon a too restricted assay system. Whereas, Table 4-2 lists the carrot assay system as a means of detecting compounds with cytokinin-like activity, it is nevertheless adaptable to

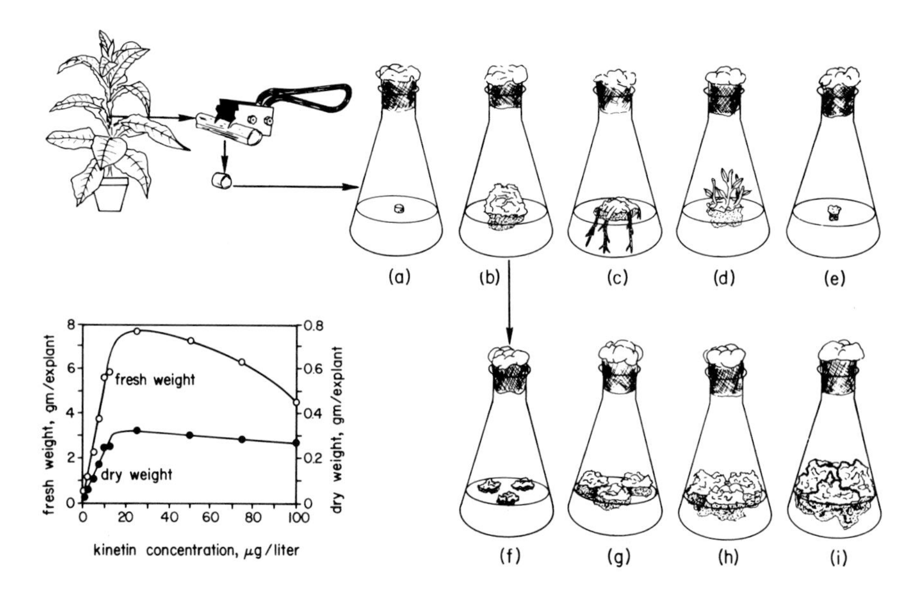

Fig. 4-7. Diagrammatic representation of the tobacco pith bioassay specially designed for testing cytokinin activity. Plants of *N. tabacum* var. Havana "Wisconsin 38" are used as a source of pith explants. Aseptically isolated tissue may be "planted" on a basal medium containing 2.0 mg/liter of indoleacetic acid and 0.20 mg/liter of kinetin. Without such additions no growth occurs (a); in the presence of auxin alone, there is slight growth (e). In the presence of kinetin (or other cytokinins), however, a vigorous callus is obtained (b). Stock cultures so obtained are then subcultured (about every 4 weeks) on a medium containing 2.0 mg/liter indoleacetic acid, but a reduced level of kinetin (0.03 mg/liter) prior to use in the bioassay. Three pieces of callus (ca. 50 mg each) are planted on test media. The series (f), (g), (h), and (i) shown here represents an increasing effect of cytokinin on growth (see the graph showing fresh and dry weights after a 5 week growth period). This assay also offers an opportunity to relate morphogenetic effects to substances. In (c) roots have developed on a medium containing 3 mg/liter of indoleacetic acid and 0.02 mg/liter of kinetin, while shoots have developed in (d) on a medium containing 0.03 mg/liter of indoleacetic acid and 1 mg/liter of kinetin. [Adapted from work of Skoog and co-workers (Skoog *et al.*, 1967; Helgeson *et al.*, 1969).]

detect and measure physiological activities of many other kinds. But there is no overriding virtue in the carrot root as such, except that of convenience. In principle, other angiosperm cultures based on totipotent cells could be adapted to similar ends.

The early users of plant tissue cultures adopted semi-solid media and relatively large tissue explants (Gautheret, 1942). These methods yielded many randomly proliferating callus cultures, with potentially indefinite growth, but which, nevertheless, grew at relatively slow rates (White, 1941, 1946). At that time the investigators were content with what they construed as "defined media" (White, 1941, 1946); the same applied in the continuous

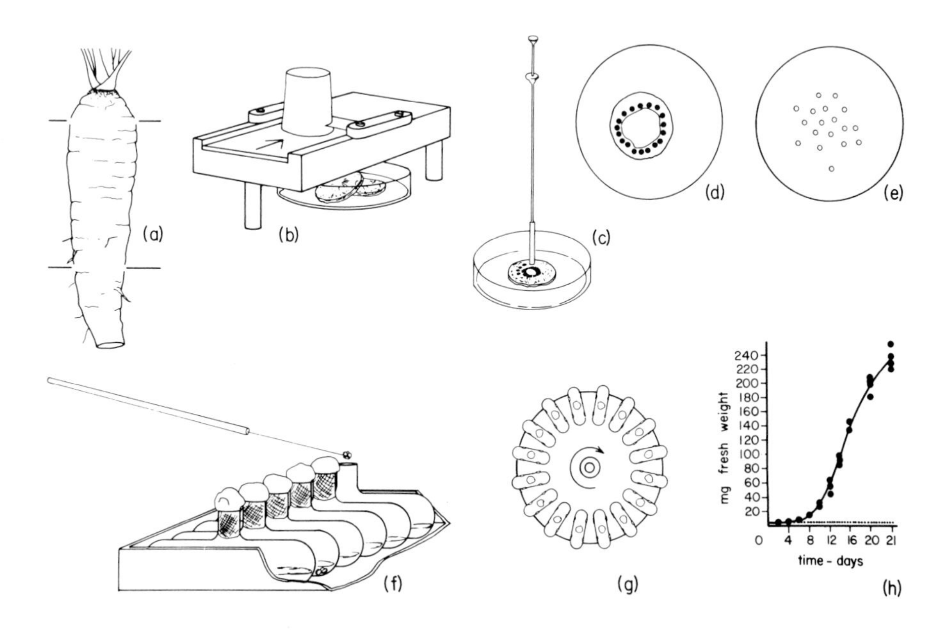

Fig. 4-8. Diagrammatic representation of the carrot root phloem bioassay. Mature roots of cultivated carrot (*Daucus carota* var. *sativus*) (a) are peeled and surface sterilized. Thin slices (1–2 mm) are then obtained using a special cutter (b). Secondary phloem explants are then removed from a distance sufficiently far from the cambium (ca. 1–2 mm) with a cannula (c and d). Explants weigh about 2.5–3 mg (e). After inoculating three explants into a culture tube, containing 10 ml of medium (f), the tubes are placed on a wheel which turns on a horizontal axis at 1 rpm (g). This rotation aerates and subjects the explants to alternate periods of exposure to medium and to air. After a specified time (e.g., approximately 18 to 21 days) the explants are removed, weighed, macerated, and cells are counted (h). (To convert milligrams to cells, 1 mg is approximately 10,000 cells.) This system allows one also to test the effects of plant growth substances and their inhibitors on such parameters as protein nitrogen content and other biochemical criteria. (From work of Steward and co-workers.)

subculture of root tips in liquid media (Street, 1957). Such media may have enjoyed a measure of definition, so far as the investigators were concerned

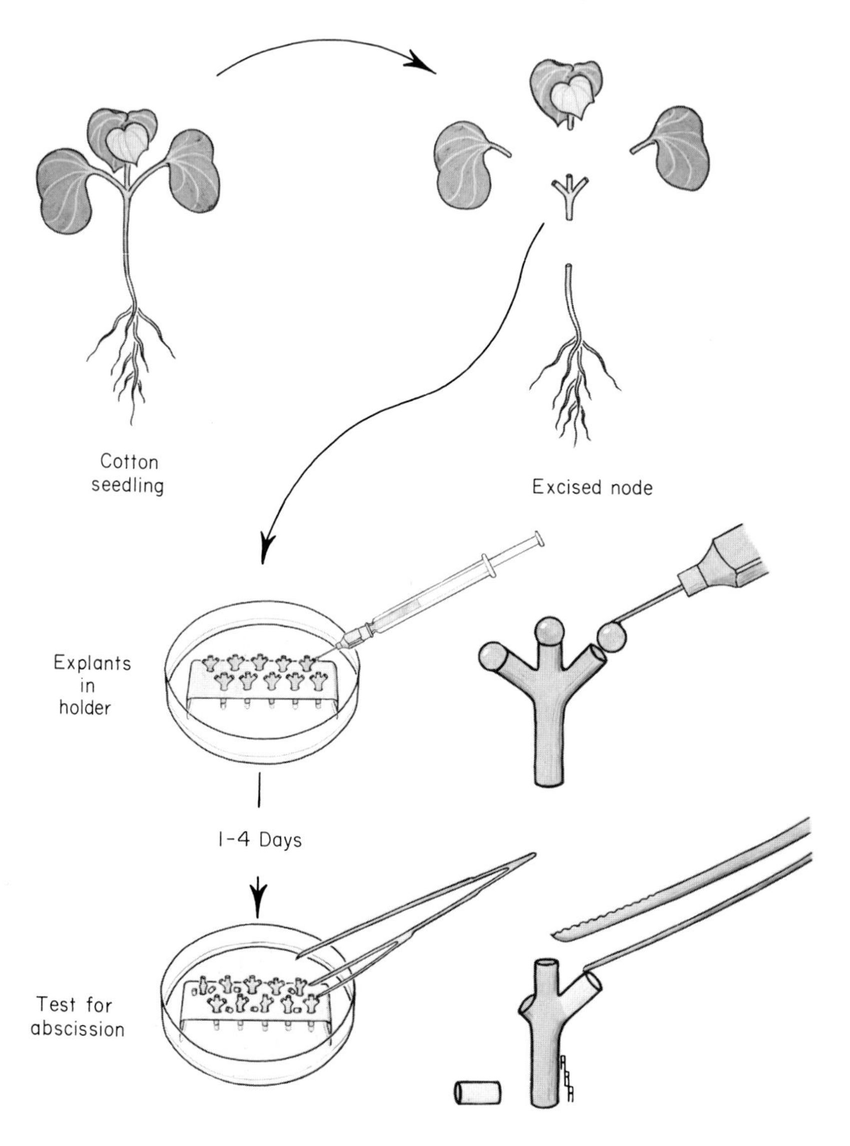

Fig. 4-9. The cotton abscission assay. The first (cotyledonary) nodes from cotton seedlings are isolated as shown. The test substance is supplied in a dilute agar solution to the exposed petiolar and stem surfaces. The test depends on accelerating or retarding the onset of abscission; the petiolar stumps abscise, or not, when "tapped" in a controlled way with special forceps. [Photograph supplied by Dr. F. T. Addicott, University of California, Davis (Addicott *et al.*, 1964).]

(see Gautheret, 1959); often they were very far from being defined, insofar as the attributes of the plant materials were concerned. Only later were the tissue isolates reduced to their minimum size to produce the maximum relative growth rates (Steward *et al.*, 1952). The growth then became dependent to the maximum extent on extrinsic factors and the range of responses induced and observed became broadened. Thus, the earlier media, as defined, became useful only as a base from which to study the beneficial effects of added stimulants. At first, these additions were necessarily complex (Steward and Caplin, 1952b) and they were derived from natural sources which had some morphological rationale (Steward and Shantz, 1959).

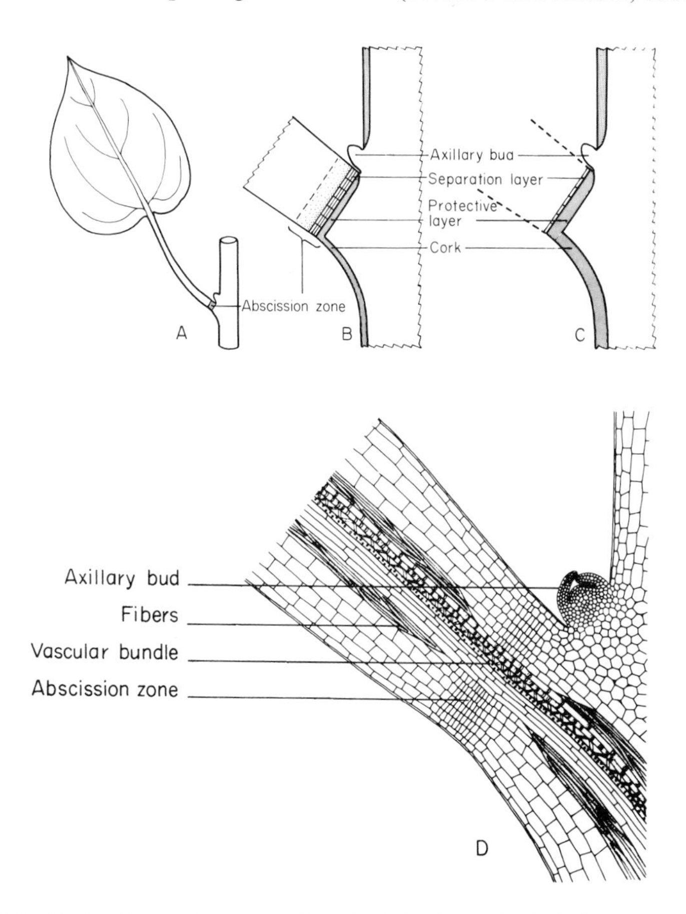

Fig. 4-10. The location and structure of an abscission layer. A–C, the location and activity of the layer in a petiole; D, anatomical detail of the abscission layer. Note the small cells and the absence of fibrous tissue in the abscission zone. It is in this preformed zone that activity may be induced, either by environmental effects or by the application of growth regulatory substances. (Diagrams supplied by Dr. F. T. Addicott, University of California, Davis.)

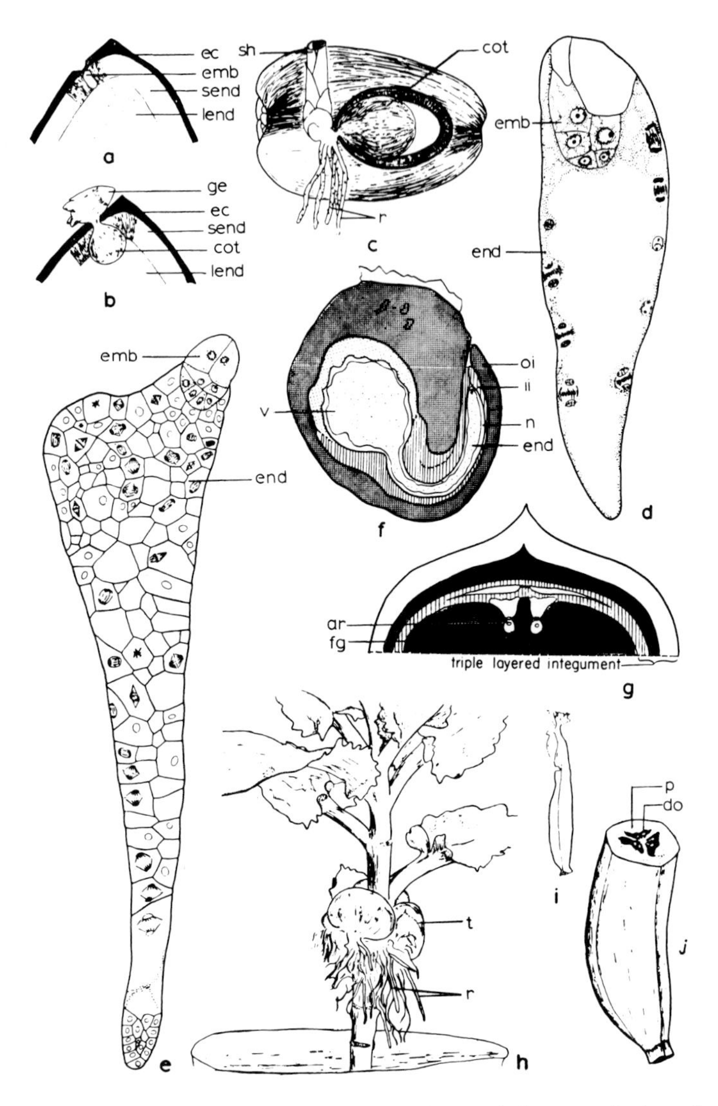

Fig. 4-11. The sources of growth factors and their morphology. *ar*, Archegonium; *cot*, cotyledon; *do*, degenerated ovule; *ec*, endocarp; *emb*, embryo; *end*, endosperm; *fg*, female gametophyte; *ge*, germinating embryo; *ii*, inner integument; *lend*, liquid endosperm; *n*, nucellus; *oi*, outer integument; *p*, pulp; *r*, roots; *send*, solid endosperm; *sh*, shoot; *t*, crown gall tumor; *v*, vesicle. (*a*) L. S. fruit of *Cocos nucifera*; note the embryo embedded in solid endosperm; (*b*, *c*); as (*a*), showing different stages of seed germination; (*d*) free nuclear endosperm in *Zea mays*; (*e*) cellular endosperm in *Zea mays*; (*f*) L. S. ovule of *Aesculus* showing free nuclear endosperm and vesicle at the chalazal end; (*g*) L. S. ovule of *Ginkgo*; (*h*) crown gall tumor on the stem of *Kalanchoe*; (*i*) the flower of banana (*Musa* sp.) at the time of shooting; (*j*) parthenocarpic fruit of the same cut to illustrate the inner layers of pericarp which develop into pulp (Steward and Mohan Ram, 1961).

Natural Sources of Growth Stimulatory Substances

Angiosperms begin as free zygotes immersed in a fluid (the contents of the embryo sac or liquid endosperm) (Wardlaw, 1965a). The first signs of the organization of dicotyledenous embryos is the transition of the globular, embryonic form to the heart-shaped condition when cotyledonary primordia develop. But the growth of the cotyledons is like that of no other leafy appendage because it first occurs when it is submerged in liquid in the ovule (Wardlaw, 1965a). But the often thick, fleshy, storage cotyledons [e.g., as in *Juglans* (walnut) or *Aesculus* (horsechestnut)] or the sometimes thin, papery ones [e.g., as in *Fagopyrum* (buckwheat) or *Ricinus* (castor bean)] both absorb and eventually transmit the contents of the medium with which they are bathed to the growing plantlet (Steward and Mohan Ram, 1961; Steward and Shantz, 1959). Therefore, cotyledons may transmit to the growing regions of shoot and root special stimuli that they elaborate. In a monocotyledonous plant, the same role is played by the scutellum, as in a cereal grain (see Fig. 2-6). Plants which shed their immature "seeds" early and which, therefore, lack both endosperm or storage cotyledons use other devices. Those that are semiparasitic upon other angiosperms, e.g., *Monotropa*, *Orobanche*, etc., may obtain both their nutrients and stimuli from the substratum and the host plant to which they attach. Those that are epiphytes, like many orchids, and which turn green, elaborate a storage organ, the protocorm, which fulfills the nutritional and metabolic role of the cotyledons or of the endosperm. By contrast, viviparous embryos, e.g., mangrove, *Rhizophora mangle*, etc., which germinate while still attached to the parent plant, receive both nutrients and stimuli from the parent sporophyte (Steward, 1968).

In the early quest for stimuli that would make quiescent cells grow again at their maximum rate, the use of liquid endosperms was tried (Caplin and Steward, 1948). By intervening prior to the full development of the cotyledons one effectively gains access to the stimuli that embryos receive via these organs. Amongst the fluids used in this way were extracts of immature corn (*Zea*) grains, as well as the liquid endosperm from the coconut (*Cocos*), from immature walnut (*Juglans*), or horsechestnut (*Aesculus*) fruits (Shantz, 1966). All these sources nourish immature embryos and are drawn from morphological situations which are in this respect similar (see Fig. 4-11). It is the philosophy to which this work has led, more than its detail (Shantz and Steward, 1964, 1968), which is now to be summarized even though there are still other morphological sources of special nutrients and of stimuli to be explored, for the growth-regulating stimuli they furnish in the division of labor between the organs of the plant body.

CHAPTER 5 Some Growth Regulatory Systems

The growth induction of otherwise mature cells, as a response not to a single substance but to a complex of interacting ones, may be dealt with in terms of the responses of small explants of the secondary phloem from the carrot root. The response of the explants from the same root is remarkably uniform when small explants are removed (between the usually tetrarch branched roots) and at a constant distance from the cambium (2–3 mm, see Fig. 5-1). Standard explants, cut in this way from the same root, constitute a clone. But even individual roots from the same variety and the same harvested stock display some idiosyncracies; these properties of individual carrot roots are obviously due to the many factors that affect the growth and development of the individual root in its own "micro" environment. To overcome this variability between roots, varieties, and seasonal crops, comparisons have habitually been made within single clones (Degani and Steward, 1968). Commonly, the entire range of treatments used in a given experiment has been replicated with different clones to show that they respond similarly, in kind, though not necessarily identically, in degree. While these variables greatly complicated the early use of the carrot assay system (Caplin and Steward, 1952; Steward *et al.*, 1952; Steward and Caplin, 1952a,b), they have now been turned to useful ends, for it is known that the different carrot clones respond differentially by growth to various partial components of the system that elicits their maximum response (Steward and Degani, 1969).

Fig. 5-1. Technique of aseptic isolation of tissue explants from secondary phloem of carrot roots (see Fig. 4-8 for further details of their use).

Synergisms and Interactions

A basal nutrient medium supplemented with 5–10% by volume of coconut milk, and with casein hydrolyzate as a general source of reduced nitrogen, elicits the maximum known growth of carrot explants. Within this range, certain partial systems, now designated I and II,* elicit part of that growth (Degani and Steward, 1968; Steward, 1970; Steward and Degani, 1969). These systems are mediated, respectively, by inositol on the one hand (System I) and by indoleacetic acid (IAA) on the other (System II). Each of these exogenous cofactors may bring into play its own range of "active factors" which at low concentration (order of a few parts per million) induce growth by cell division. These developments flow from evidence that is conveniently indicated in Figs. 5-2 to 5-6 and Table 5-1. A given

*The terminology of growth-promoting Systems I and II should not be confused with photosynthetic Systems I and II (Boardman, 1968; Duysens and Amesz, 1962), or with phases I and II in salt uptake by cells (Steward and Sutcliffe, 1959).

clone of carrot explants may respond in different degrees to the exogenous use of one or the other of these systems, depending upon the nature and severity of the endogenous limitations to its growth. A first clue to the behavior of a given clone may be its sensitivity to inositol or to IAA. But in the general use of the complex media from the vicinity of developing embryos (Steward and Shantz, 1959), these growth-promoting agents are all sup-

TABLE 5-1

CONTRASTING RESPONSES OF CLONES OF CARROT EXPLANTS[a,b,c] TO DIFFERENT GROWTH-PROMOTING SYSTEMS IN PRESENCE OF CASEIN HYDROLYZATE

		Treatments			
Clone	Growth after 18 days	B+CH	System I B+CH+Inos+AF	System II B+CH+IAA+Z	B+CH+CM
921-A	Fresh weight per explant (mg)	11.3	27.1	43.6	129.4
	No. of cells per explant (thousands)	72	772	1662	1662
	Average cell size (mμg/cell)	115	34	25	77
921-B	Fresh weight per explant (mg)	21.4	76.4	36.2	170.4
	No. of cells per explant (thousands)	287	2036	987	2279
	Average cell size (mμg/cell)	75	34	32	78

[a] Explants 921-A and 921-B, initially about 2.5 mg fresh weight and 30,000 cells.

[b] B, basal medium modified from White; CH, casein hydrolyzate (200 ppm); CM, coconut milk at 10% by volume; Inos, inositol; AF, "active fraction," i.e., ethyl acetate extract of the vesicular fluid of *Aesculus woerlitzensis*. System I, Inos at 25 ppm; AF at 2 ppm. System II, IAA at 0.5 ppm; zeatin (Z) at 0.1 ppm.

[c] *Note*: The balanced system of CM + CH produces the greatest total growth by the combined multiplication and enlargement of cells (the growth of the basal medium alone being minimal, but its relative emphasis on the number or size of the cells then being a function of the endogenous properties of the clone in question).

In clone 921-A, the medium fortified with System II produced as many cells as when stimulated by coconut milk, but these were very much smaller in size.

In clone 921-B, the medium fortified with System I produced virtually as many cells as when stimulated by the coconut milk, but these were not as large, nor were the explants as heavy.

When applied together, the two systems do not seem to be either markedly competitive or synergistic in their effects on carrot explants.

Individual carrot clones, therefore, vary in their relative responses to Systems I or II, but invariably they all respond well to the still not fully understood total stimulus of the coconut milk.

plied, although not necessarily in the same balance of Systems I and II or with the same substances that comprise their respective "active fractions."

The components of the active fraction of System II may be substituted adenines [natural ones like zeatin or synthetic ones like kinetin or many other analogous substances which have been synthesized (Shaw *et al.*, 1971; Steward, 1970)]. The active fraction of System I, however, comprises compounds which are not adenyl compounds. The best characterized member of this latter group is a rhamnose–glucose–IAA glycosidic complex (Shantz and Steward, 1968). In no case yet tried in this laboratory has zeatin, even in the presence of IAA, elicited the *maximum* growth of carrot explants such as that obtained by the use of the coconut liquid endosperm and casein hydrolyzate. Also, preparations of the complex active fractions of the native fluids (as, e.g., from *Aesculus*) have displayed great activity, even though, in comparative tests by thin-layer chromatography, they failed to show the presence of zeatin in the detectable amounts which should adequately elicit the full growth that this compound will produce (Degani and Steward, 1968).

Shaw has synthesized a series of *n*-alkylaminopurines in which the length of the side chain increased from 1 to 10 carbon atoms (Shaw *et al.*, 1971). When these compounds were added to a medium that had all other known requirements (the basal medium, B; plus casein hydrolyzate, CH; plus

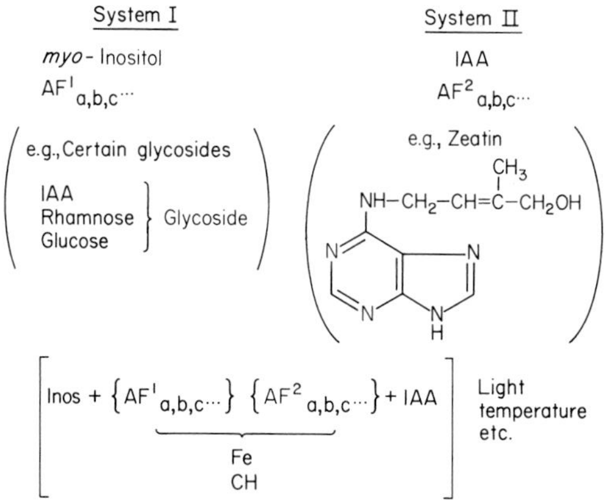

System I
myo-Inositol
$AF^1_{a,b,c\cdots}$
e.g., Certain glycosides
IAA
Rhamnose
Glucose
Glycoside

System II
IAA
$AF^2_{a,b,c\cdots}$
e.g., Zeatin
CH_3
$NH-CH_2-CH=C-CH_2OH$
N
N
N
N
H

Inos + $\{AF^1_{a,b,c\cdots}\}$ $\{AF^2_{a,b,c\cdots}\}$ + IAA
Fe
CH
Light
temperature
etc.

Fig. 5-2. Scheme to illustrate growth-promoting interactions in carrot explants. Inositol mediates the activity of several factors that induce growth (AF^1), and this interaction is referred to as System I. IAA mediates the activity of several other factors that induce growth (AF^2), and this interaction is referred to as System II. Iron (Fe) is essential for all growth induction; casein hydrolyzate (CH) extends the range of both systems. Growth is further modulated by light and temperature; certain compounds such as IAA–glycosides function as AF^1; other compounds (e.g., adenyl derivatives) function as AF^2. [From Steward (1970); by permission of The Royal Society, London.]

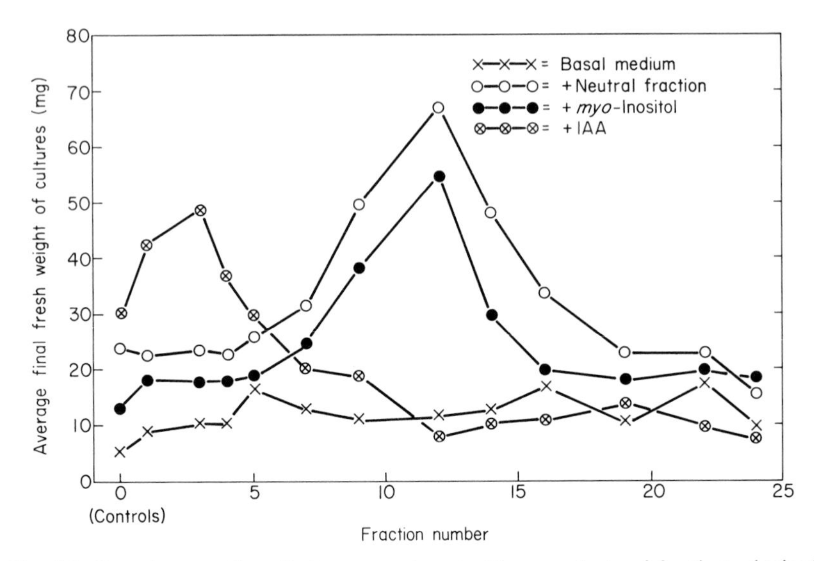

Fig. 5-3. Growth-promoting effects on carrot root phloem explants of fractions obtained from the acetic acid eluate of charcoal-adsorbed corn extract by countercurrent partition between butanol and water. Fractions were tested at 5.0 ppm in basal medium alone and in combination with corn neutral fraction (250 ppm), *myo*-inositol (25 ppm), and indoleacetic acid (0.5 ppm). Original weight of explants, 3.0 mg; growth period, 18 days. [From Shantz and Steward (1964).]

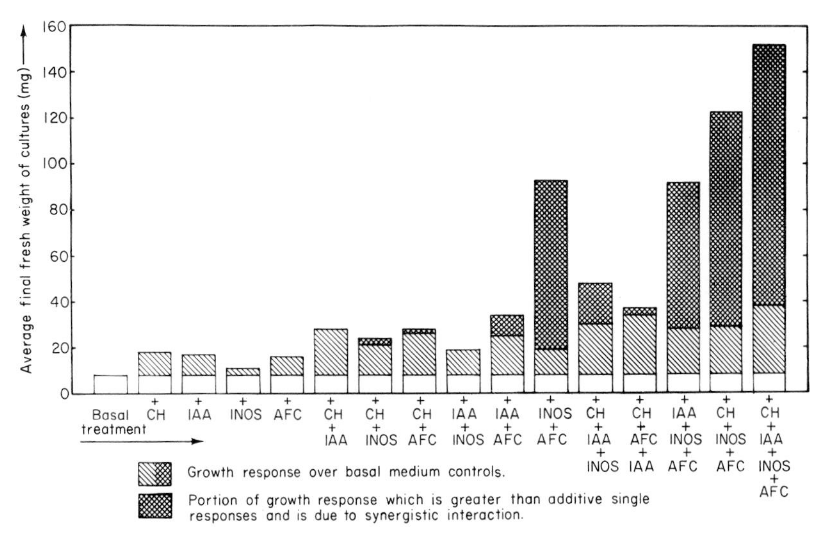

Fig. 5-4. Growth-promoting effects and interactions of casein hydrolyzate, indoleacetic acid, *myo*-inositol, and *Aesculus*-active fraction concentrate on carrot phloem explants. CH, Casein hydrolyzate (250 ppm); IAA, indoleacetic acid (0.5 ppm); INOS, *myo*-inositol (25 ppm); AFC, *Aesculus*-active fraction concentrate (5.0 ppm). [From Shantz and Steward (1964).]

indoleacetic acid, IAA; plus inositol), growth was affected and was measured after some 18 days by the mean fresh weight of the explants which grew, and by the number of cells per explant determined after a procedure of maceration. The increments due to these "adenyl cytokinins," interacting with IAA and over and above the effect of the test medium (B + CH + IAA + Inos), are shown in Fig. 5-7 for two clones of carrot explants. In each case, the level of cell multiplication attributable to zeatin is shown. Clearly, the compounds with *n*-alkyl side chains of 5 carbon atoms produce the maximum effect and, insofar as cell multiplication is concerned, the compound with a C-5 substituent was even somewhat superior to zeatin. The impressive feature is that the best ($N = 5$) of the *n*-alkylaminopurines, together with IAA and casein hydrolyzate in a basal medium, rivals the coconut milk stimulus by the number of cells it produces, but not by the size of the cells so produced, their chlorophyll content, or plastid development. In fact, when a C-5 alkylamino side chain is variously modified with CH_3, C=C, C≡C, or other (OH) substituent groups, the effect upon the cell divisions that ensues is small, but the effect on the weight of the tissue is much larger because the different compounds vary in their tolerance for, or active stimulation of, cell enlargement. Therefore, in considering the relative roles of different growth-promoting systems, it is important to keep in mind their respective effects on cell multiplication while cells remain small and on cell enlargement. The peculiar role of coconut milk and other similar fluids is that it preserves a certain balance between these two aspects of growth, even as it also modulates the formation of green plastids and maintains a characteristic metabolism in the cells.

Thus, there is still much more to be learned about the full complement of growth-inducing compounds and systems from the area of the developing embryo of angiosperms (Steward *et al.*, 1964a) and of their ability to induce growth in otherwise quiescent cells.

It is known that other compounds, natural and synthetic, can also contribute to the overall growth stimulus. Activities that are not yet integrated into the pattern of growth response that is elicited by Systems I and II, and as they interact with trace elements, with light and with casein hydrolyzate (Steward *et al.*, 1968b; Steward and Rao, 1970), are those which are due to other classes of compounds. Examples are the naturally occurring leucoanthocyanins and various phenolic compounds (Steward and Shantz, 1956), or the activity unexpectedly attributable to symmetrical *N,N′*-diphenylurea, which first appeared in the fractionation of coconut milk (Shantz and Steward, 1955). Identification of *N,N′*-diphenylurea encouraged the synthesis of other compounds which proved to be active (Bruce and Zwar, 1966; Kefford *et al.*, 1966). The activity first attributed to 2-benzothiazolyloxyacetic acid (BTOA) was somewhat fortuitously disclosed when screening

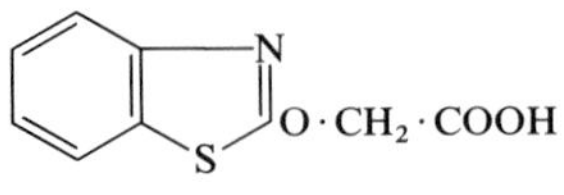

$$\text{2-benzothiazolyl–O·CH}_2\text{·COOH}$$

2-Benzothiazolyloxyacetic acid

a number of model compounds in the carrot assay test, but, although its activity is great and beyond question (Steward and Shantz, 1956), it is now open to the interpretation that its alternative, 2-oxobenzothiazolin-3-ylacetic acid, may also be involved (Brookes and Leafe, 1963). Again, un-

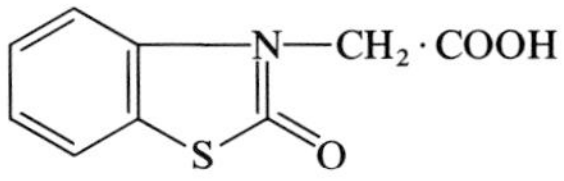

$$\text{N—CH}_2\text{·COOH (2-oxobenzothiazoline)}$$

2-Oxobenzothiazolin-3-ylacetic acid

expectedly, certain gibberellins increased carrot cell division, even in the presence of coconut milk, instead of fostering the cell enlargement that might have been anticipated (Shantz and Steward, 1964). Even the added cell division stimulus due to very low concentrations of the otherwise growth inhibitory radiolysis products of sugar is also interesting here (Holsten *et al.*, 1965; Ammirato and Steward, 1969).

The short answer is that no concise statement can yet be made which delimits the substances that stimulate cell division in quiescent cells. In some cases, single substances may suffice, more usually substances interact synergistically and in different composite systems outwardly to bring about the effects of growth. Investigators who have attributed the full growth stimulus solely to single substances did so often because the assay system used was inherently delimited in its range of response.

The different growth-promoting agents of Systems I and II modify the metabolism of the cells in very different ways, even as they affect differentially the balance between cell division and cell enlargement. For example, System II, i.e., zeatin plus IAA, tends to foster carrot cell division more than enlargement, whereas the components of System I, mediated by inositol, tend to promote more cell enlargement and more development of chloroplasts in the light (Steward, 1970). An imbalanced stimulus due to the *Aesculus* fraction and inositol of System I may lead to an excessive accumulation of nonprotein nitrogen compounds (especially of glutamine nitrogen at 75% of the total soluble nitrogen), whereas with System II (zeatin plus IAA) that accumulation is less and consists much more of asparagine and alanine (14 and 24% of the soluble nitrogen, respectively) (Steward and Rao, 1970). Somewhat similar contrasted effects in the responses of the tissue are to be seen in terms of its protein and nucleic acid content (Fig. 5-6). The range of growth responses and superimposed content of metabolites that are now attributable to the growth factor systems or their component parts (i.e., I and II), or to the most complete and balanced one known, i.e., CM, is very wide (Steward and Rao, 1970).

No single molecule has the sole prerogative of growth regulation, even within such broad classes of activity as those commonly attributed to auxins, cytokinins, or gibberellins. In this respect the history of newly observed chemical therapeutic responses has been curiously repetitive. Whether seen

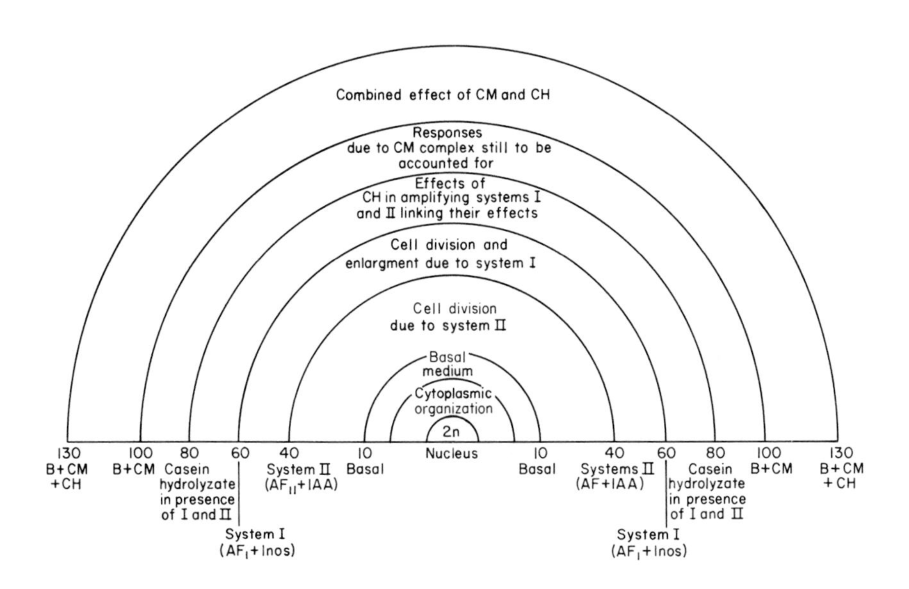

Fig. 5-5. Diagram to attribute the overall growth of carrot explants to the basal medium, casein hydrolyzate, the coconut milk complex, and its various component parts. The numbered scale represents the typical growth response in milligrams fresh weight per explant after 18 days.

B, Basal medium; CM, coconut milk (10% v); CH, casein hydrolyzate (200 ppm); AF[1], cell division factor from *Aesculus* (at 1 ppm) linked to inositol (at 25 ppm); AF[2], cell division factor linked to IAA, like zeatin (0.1 ppm); IAA, indole-3-acetic acid (0.5 ppm). [From Steward (1970); by permission of The Royal Society, London.]

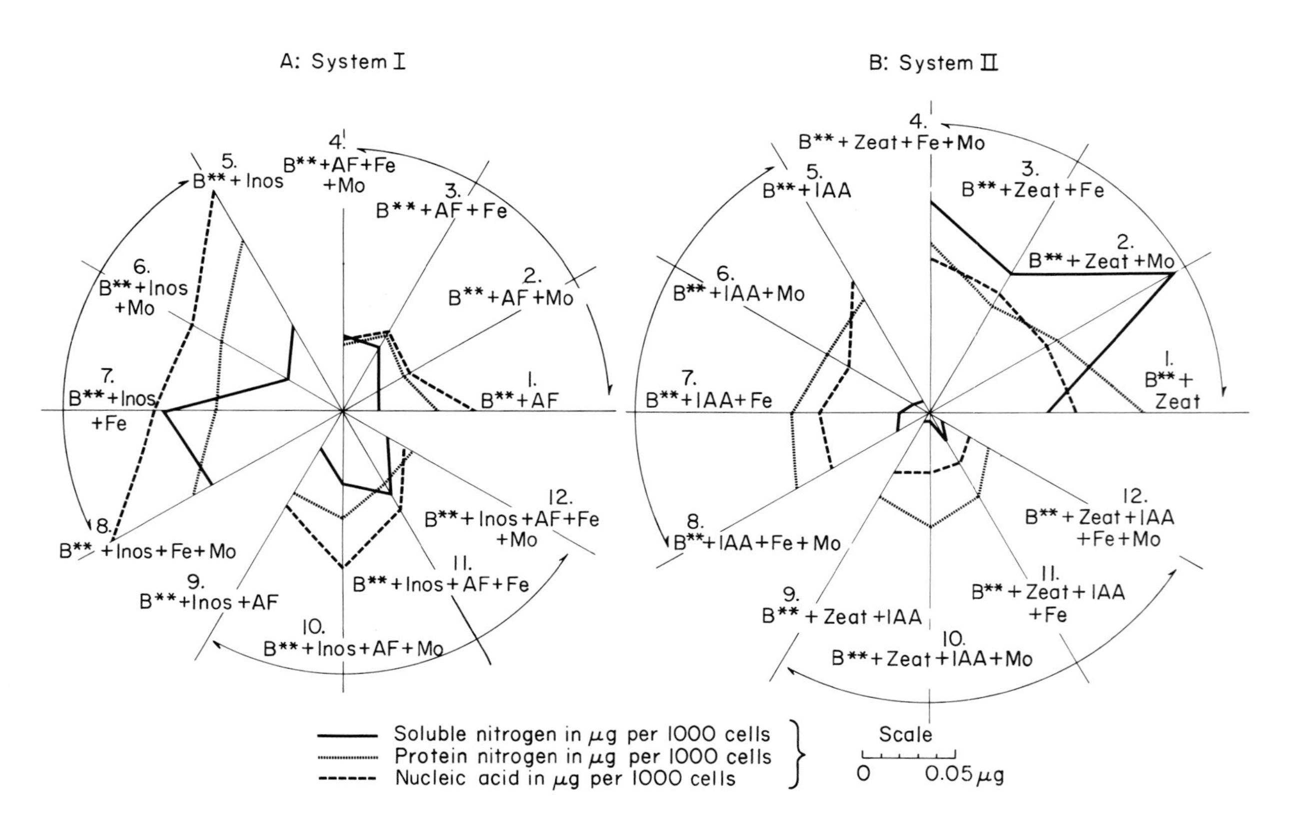

Fig. 5-6. Polygonal diagrams to show the effects of growth-promoting systems and their component parts as they interact with trace elements (Fe and Mo) and affect the behavior of carrot tissue cultured 18 days (Steward and Rao, 1970).

in terms of auxins, cytokinins, or gibberellins, the first tendency is to ascribe the supposed activity to *one* substance. This has been done with indole-3-acetic acid as *the* auxin; kinetin as *the* cell division substance; gibberellic acid as *the* first characterized substance of this class; and abscisic acid as *the* prototype of the inhibitors. Later, many similar substances came to light, and these can also qualify for a similar role in each category.

However, even if a given molecule can be identified as an overriding one in a given test for a given class of regulator, such a substance should also be considered in relation to the rest of the growth-controlling system of which it may form a part; in short, in the context of its interactions, synergistically and sequentially, with all the other parts of the total growth regulatory system. It is in these "interactions" that the most dramatic results are still to be observed.

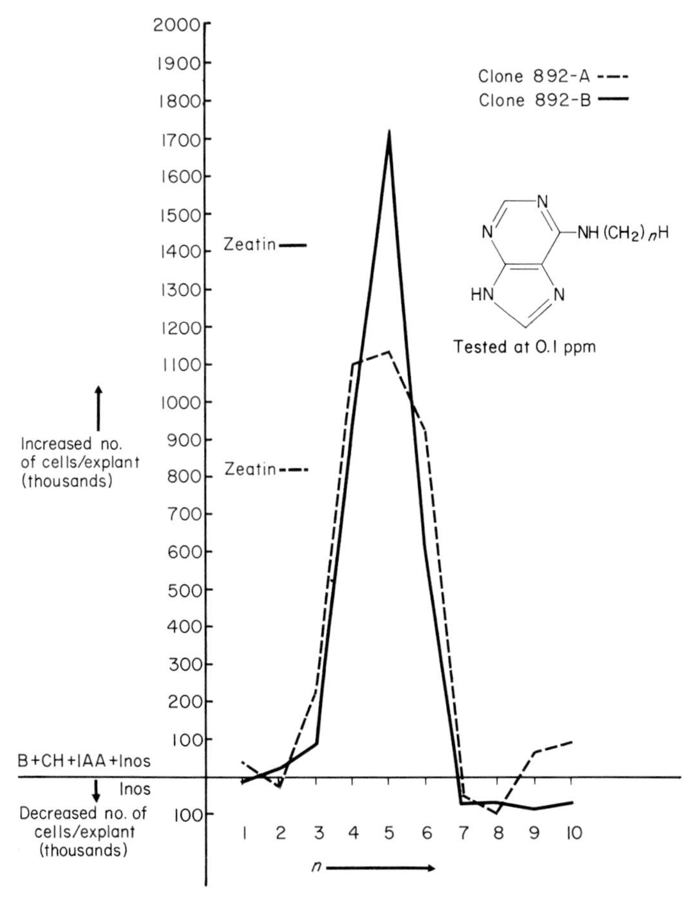

Fig. 5-7. The effect of a series of *n*-alkylaminopurines on cell multiplication in cultured carrot explants. [From Shaw *et al.*, (1971).]

Responses of Cells in Their Milieu: "Biochemical Ecology"

It is difficult to decide how far a cell needs to "ask the permission of the nucleus" for each and every act it performs, for its other organelles, e.g., plastids, mitochondria, dictyosomes, reticulum, etc., can obviously act with some degree of autonomy (Bogorad, 1967; Granick and Gibor, 1967; Whitfield and Spencer, 1968b). The problem is one of understanding how these autonomous units, once formed and each capable of a wide range of chemical reactions, have their respective activities coordinated [see Spencer and Whitfield (1967) and Tewari and Wildman (1968) for one approach to this problem]. In such complex responses as growth and development, higher orders of control may be needed. In a classic and prophetic phrase, Blackman presented in 1905 a strangely realistic concept, by viewing the cell as a "congeries of enzymes, a colloidal honeycomb of katalytic agents, as many in number as there are cell-functions, and each capable of being isolated and made to do its particular work alone *in vitro*, we look for light on the action of chemical stimuli in the cell to their effect on the action of isolated enzymes *in vitro*. Here, too, law and order is now known to reign, and while enzymes only 'accelerate' reactions without being incorporated in their end products, yet the acceleration produced is proportional to the mass of the enzyme present, minute as it is, and the effects of 'activators' are 'paralyzators' of this action are also in proportion to their masses" (Blackman, 1905, p. 294).

Such a system may be subject at any one time and place to a matrix of interlocking and interacting factors and control mechanisms. Similarities of response may be triggered off by deforming the "network," or perturbing the matrix, analogous perhaps to disturbing the "epigenetic landscape" in a variety of very different ways (Waddington, 1957). In all these concepts, the dilemma is that in higher plants the evidence of continuing communication between nucleus and cytoplasmic sites via messenger RNA's is tenuous (Key, 1969; Key and Ingle, 1968; O'Brien *et al.*, 1968; Trewavas, 1968a,b). It seems as if the morphological setting in which cells exist operates controls which determine the genetically feasible biochemical events that are practiced, rather than the reverse process (Israel *et al.*, 1969; Krikorian and Steward, 1969). If this is so, the direct involvement of the nucleus could be over when the range of information needed by each autonomous organelle is assigned; thenceforward the essential "messages" and "messengers" are exogenous substances and environment stimuli (Bellamy, 1966; Gayler and Glasziou, 1969; Nitsch, 1963; Williams and Novelli, 1968). Since such views tolerate multiple sites of action, they raise obvious complications, but they also present a challenge to those who seek to achieve chemical control over the forces of growth. As in the case of carcinogens that determine irregular and unregulated growth, we already have an embarrassing number of normal plant growth regulatory substances. But their very number and range may permit growth and other attributes of cells to be regulated in a great many distinct and highly specific ways that could help to solve the problem of the intensely local control of what cells do during development. And last, there is here a message for those who "screen" vast numbers of compounds for their plant growth-regulating attributes. It is the unusual case, perhaps too rare to be economically found by chance, in which a broad area of response can be profitably regulated solely by a single, unique substance. By contrast, the chances of achieving dramatic and useful results are far greater when regulatory substances are used in well-chosen, multiple combinations, rather than when used singly. Furthermore, activity may often be disclosed when regulators are used as part of planned sequences which affect the cells *seriatim* (Steward *et al.*, 1967).

A difficult problem, about which more evidence is needed, concerns the morphological and biochemical totipotency of cultured angiosperm cells, on the one hand, and the autotrophic nutrition of green plants on the other. But does this imply that even green, cultured angiosperm cells are individually autotrophic (Bergmann, 1967; 1968)? Or is the recognized autotrophy of the angiosperm a property of its organization and division of labor? And, if so, what does this imply for the extra nutritional stimuli which free green cells receive, so that they may exercise their full capacity to grow?

In this laboratory carrot cells may be grown free, or as proliferated cultures around explants, in a green state with normal chloroplasts in the light or in a creamy white condition in the dark, with plastids that have only a minimum of internal organization (Israel and Steward, 1966). But all of these forms are obtained under the highly heterotrophic nutritional conditions described, and the fully green, normal, chloroplasts require a full complement of exogenous growth factors for their inception. It seems, however, that once the chloroplasts have formed, the green cells should continue their growth even in an inorganic medium at virtually their full rate. So far as present evidence for carrot cells goes, this is not so. If light inocula are made, transferring cells and the minimum of the original medium in serial transfers to an inorganic medium, the growth rates of the surviving green cells fall virtually to nil. There is the obvious possibility that the physiological activity and growth of green leaf cells *in situ* is a function of their location, for they may there receive, not merely elaborated nutrients, e.g., reduced nitrogen compounds, but also stimuli from the other organs of the plant body.

Thus the understanding of the proliferating growth of cells has led to concepts of balanced and linked growth-regulating systems. It remains to apply such concepts to the interpretation of morphogenesis, i.e., the emergence of form as free cells grow.

CHAPTER 6 *Growth-Regulating Effects in Free Cell Systems: Morphogenesis*

Organized and Unorganized Development of Cultured Cells and Tissues

After the conditions for maximum growth of small aseptic tissue explants in liquid media became known, the conditions under which one could obtain viable free cells of flowering plants, and the very small cell clusters to which they give rise, followed almost in a natural sequence (Steward *et al.*, 1969). The more rapidly growing the tissue source, the more easily are viable free cells obtained. The less violent the mechanical means adopted, the more successfully are the cells isolated; the more gentle the treatments for their cultivation, the more normally do they grow (Steward *et al.*, 1970).

Dividing, freely suspended, carrot cells were obtained about 1952, and their progress by various means to multicellular clusters was observed (Steward and Shantz, 1956). The cell patterns so formed, even under uniform external conditions, appeared to be nonrandom and resembled those of early embryogeny (Steward *et al.*, 1958a,b). The free cells grew in an organized way, as in the first formation of roots, under the identical conditions that proliferating explants grew in an unorganized way. The freer cells, therefore, expressed their innate totipotency more than the same cells attached to the tissue explant. The life cycle of flowering plants may now be bridged at the level of free vegetative cells (see Fig. 6-1). This "life cycle" has been traversed many times, without any loss of vigor, by reducing the organization at each "turn" to the level of free cells (Steward *et al.*, 1964a). With appropriate sources of cells and with appropriate sequential treatments, free cells may be rapidly multiplied and from them and the small cell clusters to which they give rise, normal plantlets may be grown by the thousands,

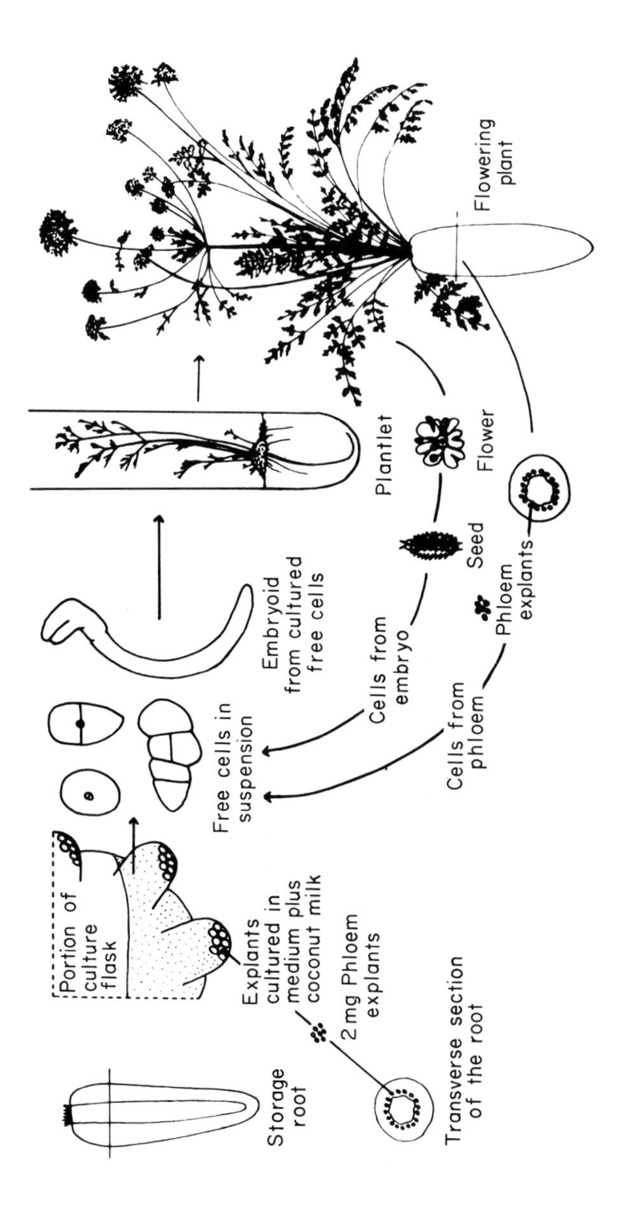

Fig. 6-1. Diagrammatic representation of the cycle of growth of the carrot plant; successive cycles of growth are linked through free, cultured cells derived from phloem or from the embryo (Steward *et al.*, 1964b).

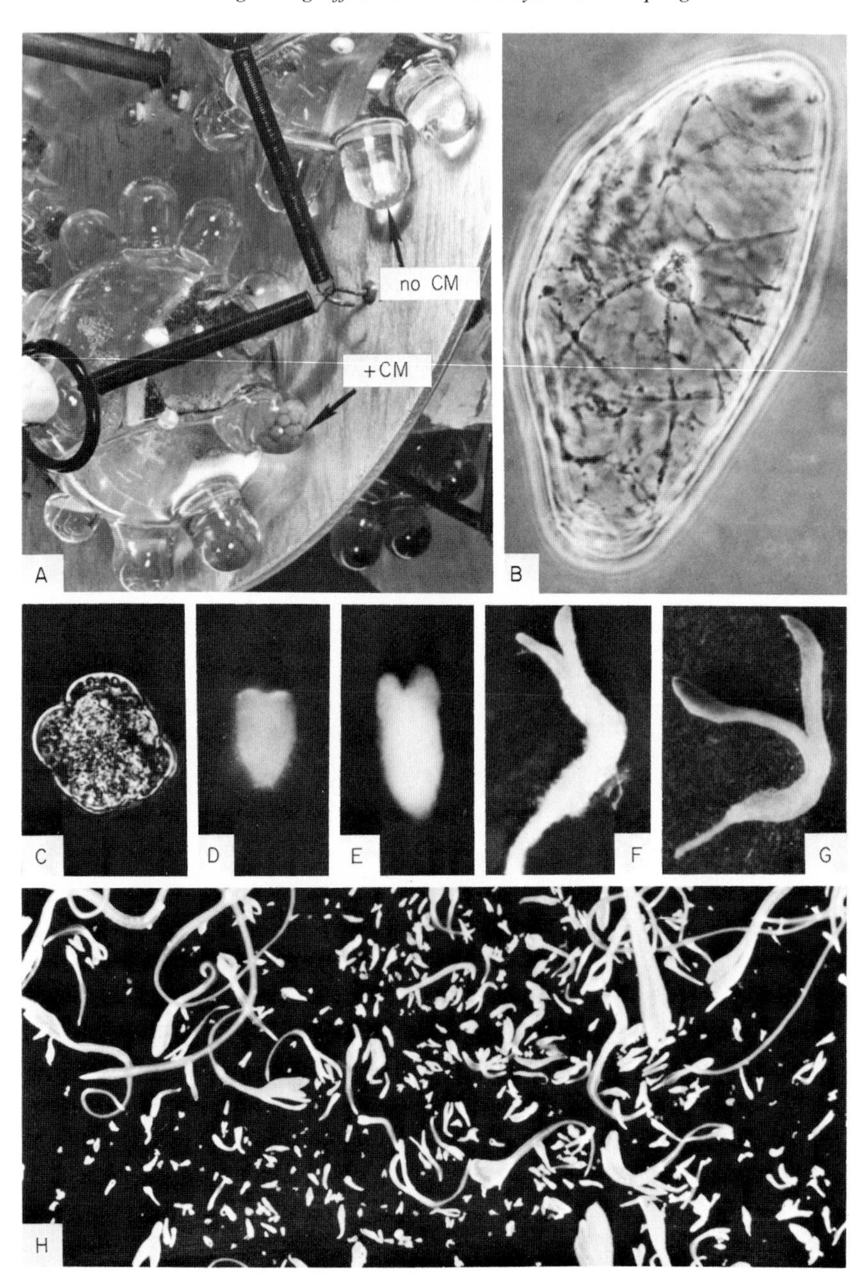

Fig. 6-2. For legend see facing page.

even in a flask which contains only 250 ml of medium (see Fig. 6-2). This approaches the long familiar propagation of fungi from conidia or asexual spores or of certain algae by aplanospores. This has been repeatedly done for carrot (*Daucus carota*), for water parsnip (*Sium suave*), and it has been shown to be feasible for certain other umbellifers (dill, anise, coriander); it has been achieved for tobacco (*Nicotiana rustica* and *N. suaveolens*) and has been accomplished with varying degrees of success with asparagus (*Asparagus officinalis*), an orchid (*Cymbidium* spp.), a crucifer (*Arabidopsis thaliana*), etc. (Steward *et al.*, 1966). In some cases, the normal development of embryos from zygotes may be recapitulated very closely, e.g., *Daucus*, *Sium*, and *Cymbidium*, but, even in these examples, the experimental conditions may so change the course of development that aberrant growth forms emerge. In other cases, the conditions promote the multiple formation of adventitious buds and regenerated roots in proliferating colonies of cells.

While a full account of the ways in which the external medium may affect the course of development from free cells has been given (Steward *et al.*, 1970), suffice it to say that the following responses of some umbelliferous plants have been observed.

The total osmotic pressure of the medium has been varied in terms of both sugars and salts, e.g., of K^+. The concentration of the medium may keep the "embryoids" small and compact, as *in situ,* or it may allow premature elongation of roots or of hypocotyls. Cotyledons, which normally become fleshy and store food materials in seeds may be bypassed, for, after they have been initiated, they may develop prematurely into expanded dark green leaves. This may occur in response to a high level of reduced nitrogen (NH_4^+ in contrast to NO_3^-) in the medium (see Fig. 6-3).

But one should point here to the futility inherent in the pretense that any named medium has, by its composition, any overriding claim to general applicability. Although such named tissue culture media are now very nu-

Fig. 6-2. Growth of carrot explants, free cells, and their embryogenesis. A, Culture flasks containing 250 ml of a basal medium, with and without coconut milk (10% by volume), showing the effect on the growth in 20 days of the original explants (2.5 mg). In presence of coconut milk the explants grew (to approximately 250 mg) and were green, in contrast with those in the basal medium which remained small and orange in color. B, A carrot cell ($300 \times 125\ \mu$) freely suspended in the coconut milk medium as it might have originated in the flask at A. C–G, Stages in embryogenesis that developed in free cell suspensions; the magnification decreases from the microscopic globular form at C to the cotyledonary stage at G. H, A random sample of a crop of carrot embryos, slightly magnified, as developed from free cells by a sequence of treatments. The field shows all stages of development from globular embryos to plantlets and, as cells are sloughed off, embryos are repeatedly formed. [From Steward (1970); by permission of The Royal Society, London.]

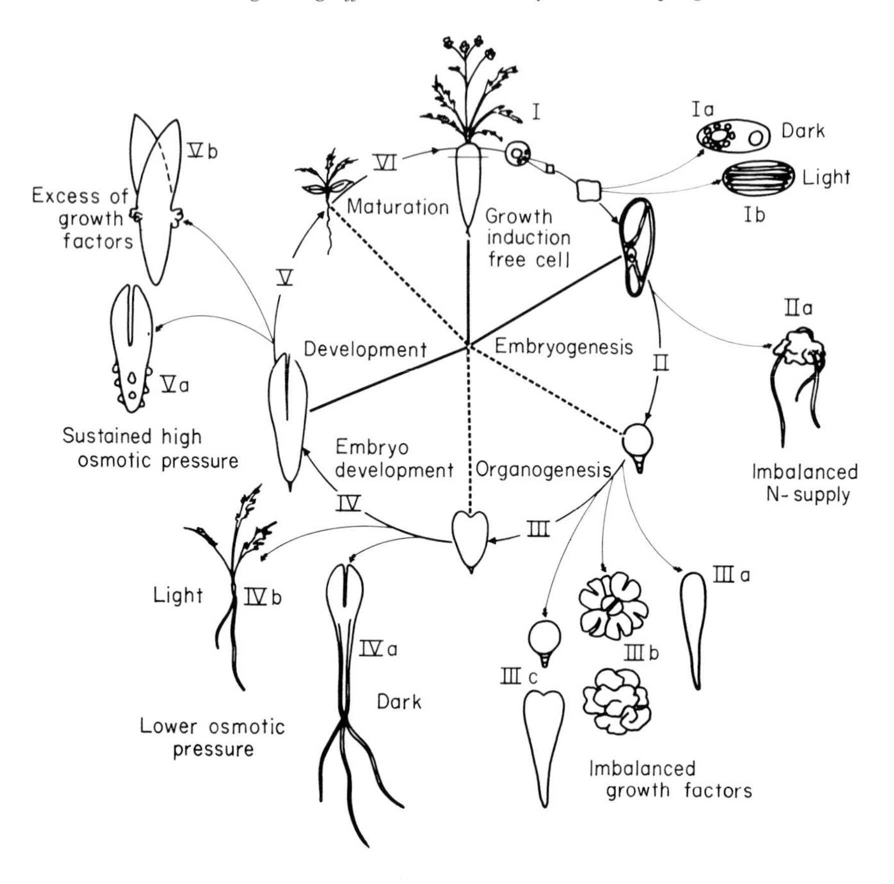

Fig. 6-3. Totipotency of cells: induction of growth and controls of development. I. *Growth induction.* Quiescent cells of storage roots are activated to grow and proliferate by a complex of factors, as in coconut milk, over and above the basal nutrients. These factors comprise cell division factors (cytokinins) which interact with inositol (System I) and with IAA (as System II), and both are accentuated by reduced nitrogen compounds. In this complete nutrient plastids reach full size in the dark but need light for morphogenesis to chloroplasts. (This growth is also a function of length of day and temperature.) Free cells formed in this way prove to be totipotent. The free cells are less tolerant of higher salt concentrations than tissue explants or cell clusters.

II. *Embryogenesis.* Isodiametric free carrot cells may simulate embryogenesis when they grow, especially in a medium which also simulates the environment of a zygote. This has often been achieved by the use of coconut milk together with naphthaleneacetic acid to foster rapid multiplication but, for later development, the auxin should be removed (as at III). This has been termed the sequential effect. In solutions limited in reduced nitrogen, roots may be formed, preferentially to shoots and in place of proembryos, as at IIa; proembryo formation and early organogenesis (as at III) are little affected by light or darkness or the osmotic pressure of the medium.

merous and are identified with White, Gautheret, Heller, Street, Reinert and Torrey, Murashige and Skoog (to mention only a few) (see Street and Henshaw, 1966), these are merely convenient general sources which prescribe, qualitatively, the main basal nutrients required by virtually all plant cells which grow heterotrophically. The requirements for both nutrients and stimuli may need to be varied, not merely from plant to plant, but from one stage of development to another. Late exposure of the plantlets to the powerful cell division producing factors of coconut milk, or other sources, may even cause localized growing centers (buds) or tumorlike calluses to form. A special causative agent of abnormal development is present in γ-irradiated (by ^{60}Co) sucrose. This substance(s) is toxic at relatively high levels, but stimulatory at lower levels when morphogenetic effects appear (Ammirato and Steward, 1969).

The organized development of cells into embryos, or into shoot or root tips, requires nonrandom patterns of cell division. The classical literature of embryology documents these patterns in detail for the zygote *in situ* (Johansen, 1950; Maheshwari, 1950; Wardlaw, 1955, 1965a). To stimulate cultured angiosperm cells into embryogeny, they should be: (a) physically, or physiologically, freed from a parent tissue mass; (b) caused to grow rapidly. Free cells can then express their innate potentialities, and the

---

III. *Organogenesis.* Although free cells of carrot may form roots readily (as at IIa), the development of normal plantlets occurs best if they form normal embryos which, in fact, initiate cotyledonary primordia. This is fostered by a plentiful, balanced, nitrogen supply and a low level of auxins or cell division factors which tend to produce abnormalities, as at IIIa (lack of cotyledonary primordia) when nitrogen is limiting; when growth factors are imbalanced, as at IIIb, multiple shoots or even calluses appear due to excessive cell division. Control of the auxin level leads to either undeveloped embryos or those with elongated radicles (as at IIIc).

IV. *Embryo development.* This can still occur in the dark. The torpedo-shaped embryos are fostered by high nitrogen supply, by relatively high osmotic concentrations which keep the embryos small but, at this point, exogenous supplies of growth factors (auxins or cell division factors) are not needed. At low levels of nutrients, hypocotyls greatly elongate (as at IVa) in the dark, but not in the light, when embryonic shoots unfold (as at IVb).

V. *Plant development.* The normal torpedo embryos develop in the light at low osmotic pressure into normal plantlets. At this stage exogenous growth factors (auxins or cell division factors) are not needed. As normal development (as at IV) occurs, the totipotency of cells is arrested *in situ* and physiological gradients are established. At high osmotic pressure, development is arrested (as at Va); with excess of growth factors adventive hypocotylary embryos (as at Vb) occur.

VI. *Maturation.* As organs mature, roots, and shoots respond to morphogenetic stimuli and, in the roots, cells enter a nongrowing, quiescent phase from which they emerge, due to the stimuli of growth induction as at I. Different degrees of stimulus are needed to induce the cells of different organs to grow. [From Steward (1970); by permission of The Royal Society, London.]

daughter cells retain these if they separate, but they express them only incompletely, or to different degrees, if they remain attached. The protoplasts of attached cells are connected and integrated by plasmodesmata, even across the first formed cell walls (see Voeller, 1964). There are often visible differences of (a) form and (b) composition as between the cells that arise even from the first divisions. In some cases this asymmetry, spontaneously established under uniform conditions, can lead to the organized formation of embryonic shoots and roots and, thereafter, to normal heart-shaped, torpedo-shaped, and eventually cotyledonary embryos (see Fig. 6-2, C–H) in great profusion. Under suitable conditions the process becomes virtually continuous, as cells which are sloughed off from the hypocotyl of the developing plants can repeat the process. But a single stimulus does not always suffice to bring cells into the conditions for morphogenesis, for a sequence of balanced stimuli may be needed. This so-called "sequential effect" was observed empirically, and in application it has proved to be a potent factor in the chemically induced morphogenesis of otherwise randomly growing cells (Nitsch, 1968; Steward *et al.*, 1967).

Asparagus cells cultured on a basal medium (B) respond by a relatively slow and unorganized growth to the addition of coconut milk (CM) and α-naphthaleneacetic acid (NAA at 5 ppm). After a period of cultivation on this medium (B + CM + NAA), and not before, their later transfer to a similar medium (B + CM) which contained 2,4-dichlorophenoxyacetic acid (2,4-D at 5 ppm) caused the cells to develop more conspicuous internal organization. By subsequent treatments, the cells so stimulated gave rise either to the formation of multiple shoots or roots or both. The point here is the following.

When free cells spontaneously recapitulate embryogeny, then there must be some form of "chemical conversation" between daughter cells which will determine their respective roles. This is the well-known origin of polarity, but it can occur under the most uniform conditions of light, temperature, aeration, and orientation to gravity that we can create (Wardlaw, 1965a). The derivatives of one cell are in some way committed to certain choices which are eventually compatible with shoot formation, whereas those of another may adopt alternatives which are compatible with roots, and so on. These alternatives do not necessarily involve any loss of inherent genetic totipotency of the individual living cells. The price paid for the specialization of higher plants, which have achieved such great complexity of form, is the restriction during development of the functions which their cells perform under conditions which are severely prescribed (Wardlaw, 1965c). To maintain the integrity of the whole, these available choices are reciprocal ones, viz., negative geotropy vs. positive geotropy; green vs. nongreen; photosynthetic carbon dioxide–reducing vs. carbohydrate-requiring; nitrate-reducing vs. reduced nitrogen–utilizing, etc. There are no single genes that control these morphogenetic events, i.e., genes for shoots or roots respectively; and the choices, made at a higher level, must involve many hereditary units acting together. If the free cells as they grow are unable to make these paired reciprocal choices simultaneously, or under uniform and constant conditions of growth, they may require the special stimulus of more prescribed conditions of external environment and of exogenous chemical regulation. These conditions may be needed to elicit appropriate forms, e.g., of shoots or roots; they may also be required to promote the development of organelles, e.g., chloroplasts, which are concerned with major features of biochemistry (photosynthesis) or even to elicit some critical, limiting reaction. The sequential treatments already described seem to have done this, empirically, in certain cases.

The Chemical Control of Unorganized Systems: The Challenge

Any discussion of the chemical control of organogenesis in otherwise unorganized callus cultures should begin with the observations of Skoog and Miller (1957). Large, proliferated cultures of tobacco stem callus were caused to form buds, or roots, by manipulating IAA and kinetin in the culture medium (see Fig. 6-4); but it is recognized that the *de novo* formation of shoot or root meristems from cells must also involve their properties as they occur, locally, within the tissue mass.

It would be gratifying if the stimulus to cause free carrot cells to form shoots or roots could be ascribed to the differential action of the growth-promoting Systems I or II, which induce cell division by different biochemical routes; or, alternatively, if it could be ascribed, in other plants, to one or the other of the sequential steps by which morphogenesis has been achieved. One can only say that if this result, tantamount to the hypothetical organ-specific "calines" of Went (*caulocaline* and *rhizocaline*, envisioned as stimulating stem and root formation respectively) (see Thimann, 1948, 1952, and references there cited) is to emerge, this is still for the future.

It would also be satisfying if the strikingly localized effects of differentiation (as in the formation of xylem and phloem elements from cambium, or of tissues behind apical meristems) could also be due to localized balanced effects of such factors as auxins and gibberellins, etc., as several workers now seem to think (DeMaggio, 1966; Jacobs, 1970; Torrey, 1966). Although such views follow along lines pioneered by Wetmore (Rier and Beslow, 1967; Wetmore and Rier, 1963), invoking the role of IAA and sugar in the formation of xylem elements in cultures of callus, any final conclusions are still for the future.

A full measure of control over the inherent biochemical and morphogenetic capacity of surviving angiosperm cells will require the study of multiple combinations of physiologically active substances applied both synergistically and sequentially in time. Even at constant conditions of light and temperature the potentialities here are great, but the logistics of experimental design become formidable. Whereas the experiments leading to the results of Fig. 6-2 are carried out aseptically under conditions of constant, controlled light and temperature, the full range of responses in any given system may also require any or all of the possible combinations of diurnally fluctuating light and temperature. For this work, controlled growth chambers are essential.

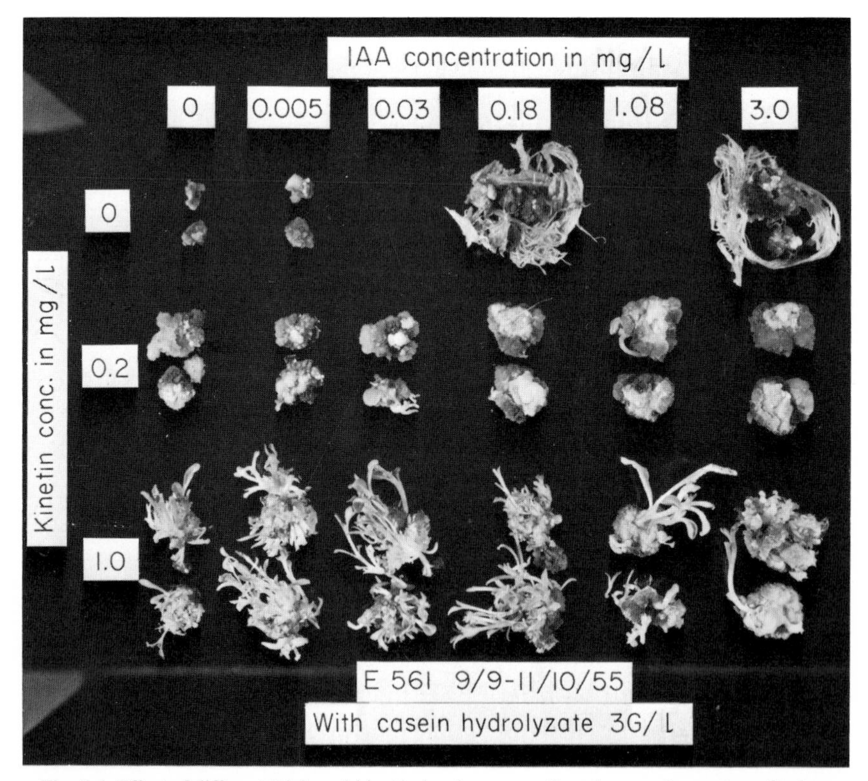

Fig. 6-4. Effect of different IAA and kinetin levels on growth and organ formation of tobacco callus cultured on modified White's medium (semisolid) plus casein hydrolyzate. [Photograph supplied by Dr. F. Skoog, University of Wisconsin (Skoog and Miller, 1957).]

When one recognizes what seems obvious—that developmental situations require a programmed chemical environment to be integrated also with diurnally controlled sequences of day length and periodicities of temperature—the complications become great. One can prophesy with confidence, on the basis of present evidence, that the potential gains are great, for they hold out the promise of:

(a) Stimulating surviving isolated cells to deliver at will their characteristic biochemical products without first creating the morphological setting, i.e., the organs of the whole plant, in which this normally occurs (Krikorian and Steward, 1969).

(b) Producing at will large numbers of plantlets in genetically cloned strains in a system of micropropagation from unorganized cultured cell units, by a technique which, if perfected, could find numerous practical applications (Steward *et al.*, 1970).

The controls over the incidence of growth are exercised by a balance of factors which (a) promote and (b) which antagonize. When these controls go astray, tumors, or abnormal forms of development, may arise (Braun, 1969a, b). The formation of cell patterns and of organs is not solely concerned with the rate, or total amount, of cell growth, which if uncoordinated would only lead to proliferation, for it also requires the controlled growth of cells as in morphogenesis. The emergence of form in otherwise unorganized cell cultures, which is rarely achieved in animal cell cultures, has, in plants, yielded to, and is obviously controllable by, manipulations of the composition of the ambient medium with which cells are bathed and of its changes with time (see Halperin, 1969, and references cited therein). But a key role in the morphogenesis of plants is the fact, more evident than in animals, that their cells are not only in static proximity, but their protoplasts are in organic connection across their cell walls. Therefore, the growth-regulating role of chemical substances is not confined to those which the cells receive via their ambient media, for they also involve those that are transmissible from cell to cell and, as in the early concept of plant hormones, from one organ to another. Whereas Thimann (1965) has likened all this to a plant's version of an endocrine system, it seems more appropriate to liken it to a system of chemical embryology which, though highly specific to plants, is analogous to that which Needham (1931) elaborated for animals.

The regularity with which organs arise (leaf primordia, as in phyllotaxis; lateral roots where they occur, etc.) testifies to the regularity with which the causal stimuli are delivered to, and received by, initiating cells. And the evident contrast in the form of the apices of vegetative and floral shoots (Wardlaw, 1965b), and in the patterns of cell growth and division as they grow, all emphasize a combined environmental and locally induced control over what the cells do where and when they are. Often, in shoot tips, there are subterminal regions in which cell divisions are scanty (Clowes, 1961); these may be formative zones which often spring into action on flowering. There is great scope here for localized chemical controls which maintain a level of quiescence until, with the onset of flowering, the appropriate stimuli (whether photoperiodic or thermally periodic, and however preceived) may mediate the onset of cell growth and cell division by transmitted chemical messages. In the sense that the still hypothetical florigens (Chailakhyan, 1968a) are truly flower-forming substances, they do not seem to exist; in the sense that they are agents which incite cells to grow in the regions and under the circumstances that flowering ensues, they are to be linked with all the other chemical agents that control the growth of cells (see Lang, 1965). Similarly, the "chemical messages" which pass from organs of perception to sites where tubers or bulbs form, and which alert quiescent cells to grow and to store nutrients, are similar to, though not necessarily identical with, the chemical stimuli which act exogenously on free cells to determine their growth.

The facts of maturity of organs, of senescence, of dormancy, and of reactivation after periods of rest, all need to be integrated into a comprehensive understanding involving the chemical regulation of cell growth and of the quiescence and reactivation of cells (Rappaport and Wolf, 1969; Roberts, 1969; Wareing, 1969b). The most important knowledge will be that which specifies the chemical agents which perform these acts naturally in the plant body. An enormous literature has developed about exogenously applied substances which profoundly modify the behavior of plants and their organs. Such substances intervene externally, and arbitrarily, in the continuing "chemical conversations" within the organism by which its development is continuously regulated.

CHAPTER 7 *The Range of Biologically Active Compounds*

This chapter surveys the range of compounds that function or are used as growth regulators. The order is arbitrary but shows the range of structures and varied actions.

Indolyl Compounds

The early workers recognized that curvatures of shoots, associated with unilateral illumination (see Thimann, 1967a) or the geotropic stimulus (see Audus, 1969), involved an influence of a chemical nature. Many naturally occurring auxins are now known (see Bentley, 1958, 1961; Pilet, 1961; Stowe, 1959), although indole-3-acetic acid (IAA) is still the most important. Other examples are indole-3-acetaldehyde (even indole-3-ethanol), β-

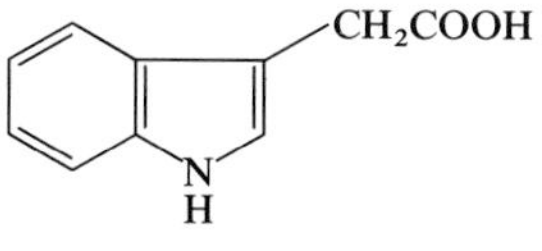

CH_2COOH

N
H

Indole-3-acetic acid (IAA)

(indole-3)- propionic acid, γ-(indole-3)-butyric acid; (indole-3)-pyruvic acid; (indole-3)-glyoxylic acid; (indole-3)-glycolic acid; (indole-3)-acetonitrile

(IAN); 1-(indole-3)-acetyl-β-D-glucose; (indole-3-acetyl)aspartic acid (Bentley, 1958, 1961). IAA–sugar and IAA–amino acid complexes are recently discovered forms of the combined auxins (Bandurski *et al.*, 1969). Glucobrassicin (3-indolylmethylglucosinolate) and its 5-methoxy deriva-

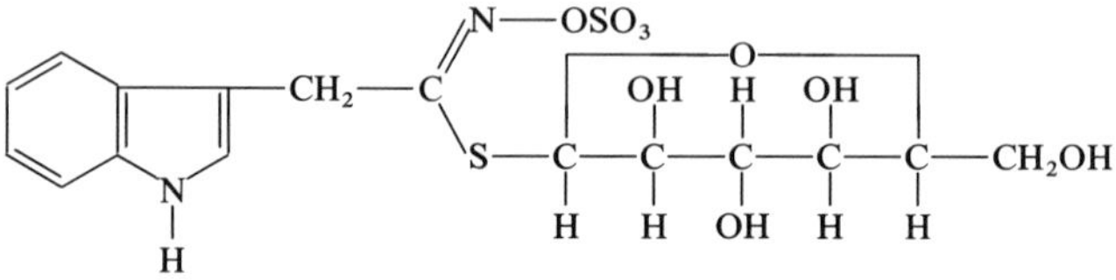

Glucobrassicin

tive (neoglucobrassicin) are bound auxins which break down to yield, among other products, ascorbigen, a bound form of ascorbic acid (Kutáček,

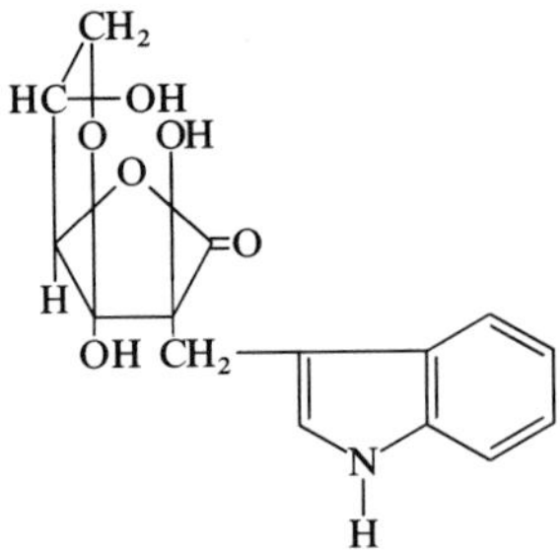

Ascorbigen

1967). Other indole compounds, e.g., isatin, indole-2,3-dione (James and Wain, 1968), have restricted occurrence (see Epstein *et al.*, 1967; Stowe, 1959, for a review of naturally occurring indoles). Still other auxins are being identified as work proceeds.

Many synthetic substances induce cell elongation similar to that produced by IAA. Some of these are more active auxins than the natural substances and induce many different responses. The best known groups of synthetic growth regulators that followed upon the discovery of auxins are indole-3-butyric acid; naphthalene compounds involving α-naphthaleneacetic acid and its derivatives; 1-naphthoxyacetic acid and its higher homologs; substituted phenoxy aliphatic acids, their esters and salts (especially 2-chlorophenoxyacetic acid; 4-chlorophenoxyacetic acid; 2,4-dichlorophenoxyacetic acid; 2,4,5-trichlorophenoxyacetic acid, and their higher homologs); and substituted benzoic acids involving 2,3,6-trichlorobenzoic acid, 2,3,5-triiodobenzoic acid, and 2,5-dichlorobenzoic acid (see Wain and

Fawcett, 1969). To these may be added unsaturated hydrocarbon gases, especially ethylene (Spencer, 1969) released by plants and certain compounds now known to yield ethylene on metabolic breakdown (Cooke and Randall, 1968; Palmer *et al.*, 1967; Pratt and Goeschl, 1969).

Adenyl Derivatives

A substance discovered by Skoog and his co-workers in autoclaved samples of DNA was shown to be very active in combination with indole-3-acetic acid in the tobacco pith bioassay (Miller *et al.*, 1955a). This substance, later identified as 6-furfurylaminopurine (a derivative of adenine; see formula for ring numbering) was named kinetin, from "kinesis," to imply

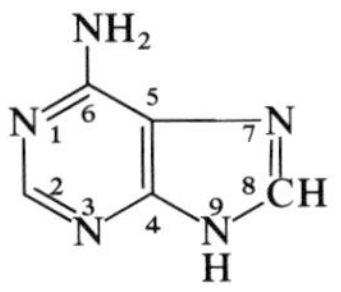

Adenine

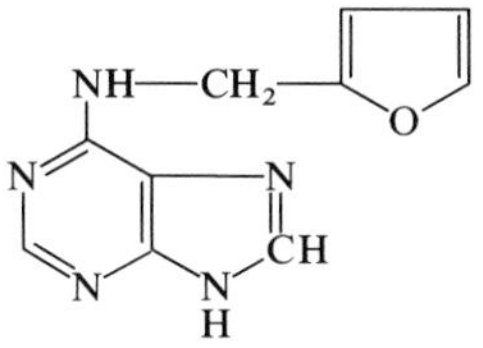

Kinetin

a substance that promotes cell division (Miller *et al.*, 1955b). From this trivial name of a substance the term kinin derives, and this referred to a generic class of substances that produce similar effects (Strong, 1958). To avoid confusion with the term "kinin" as it was previously used in animal physiology, the terms "cytokinin" or "phytokinin" emerged. In later investigations of the activity of substituted adenines and their role as cell division factors, Skoog has narrowed the definition of a cytokinin so that it must essentially be a N^6-substituted adenyl compound (see Skoog and Armstrong, 1970).

Although kinetin is generally acknowledged to be an artifact, not naturally occurring, other N^6-substituted adenine derivatives are responsible, in part, for the growth-stimulating activity found in many plant extracts (Hall and Srivastava, 1968; Letham, 1968), and these will be discussed with reference to their mode of action in Chapter 8.

Zeatin [6-(4-hydroxy-3-methylbut-*trans*-2-enylamino)purine] was the first naturally occurring adenyl cytokinin rigidly identified after isolation from extracts of sweet corn (Letham, 1967a). How closely the occurrence of zeatin is tied to the genetics of "sweet" corn seems not to be known. Other

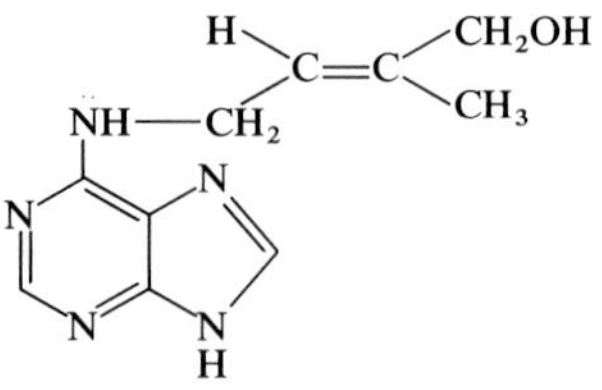

Zeatin

related cytokinins have been isolated from corn and identified as the nucleotide N^6-(*cis*-4-hydroxy-3-methylbut-2-enyl)adenine [9-β-D-ribofuranosylzeatin-5′-phosphate (Letham, 1968)]. Letham (1968) has also isolated and identified zeatin nucleotide from coconut water. Dihydrozeatin, N^6-(4-hy-

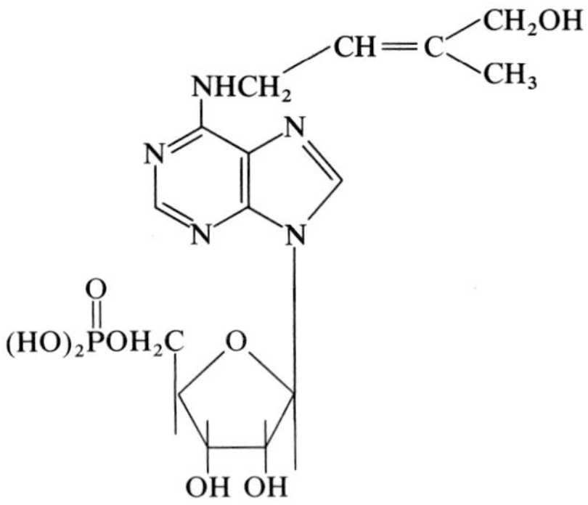

Zeatin nucleotide

droxy-3-methylbutyl)adenine [6-(4-hydroxy-3-methylbutylamino)purine], and its riboside have been isolated from immature seeds of *Lupinus* (Letham, 1968). Although zeatin and/or related compounds now appear to be naturally occurring growth regulatory substances, it should be remembered that *maximal* growth rates of cultured tissues cannot yet be achieved with defined media using zeatin as the sole prototype of the cell division factors when compared with such natural fluids as corn extract, coconut milk, *Aesculus* liquid endosperm, etc. Highly active preparations demonstrably free of zeatin have been obtained from the coconut and also from *Aesculus* (Degani and Steward, 1968; Steward and Degani, 1969). Factors over and above those like zeatin must also be invoked as responsible for the obtained growth responses. One serious objection to the view that zeatin or its relatives are solely responsible for the growth responses obtained is that very small amounts of zeatin are detectable in the natural extracts, e.g., 0.4 mg was isolated from 60 liters of coconut milk (Letham, 1968). Amounts needed for activity are far in excess of this, e.g., 10% v/v or less of coconut milk is ordinarily employed as a growth supplement (Shantz and Steward, 1964). Likewise, Miller (1963, 1968), who considers adenine to be a cytokinin, is faced with explaining why such large amounts are needed to elicit a growth response. Adenine can hardly be considered a cytokinin if small amounts are detectable *in situ* and large quantities are needed to stimulate growth in tissue culture bioassay systems. The difficulty here of course is that the high amounts of exogenous adenine may in fact react in cultured tissues to form compounds (adenyl cytokinins) that may account for the responses observed.

There are points of principle here. Although "high cytokinin activity" is by definition limited according to Skoog to N^6-monosubstituted adenyl

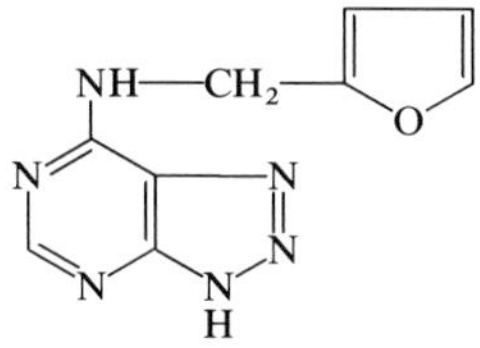

8-Azakinetin

derivatives, nevertheless, azakinetin (see Steward, 1968) is active despite the fact that it is not an adenyl compound. In other words, replacement of the number 8 carbon by nitrogen in the purine ring does not eliminate activity. Skoog himself (Skoog and Armstrong, 1970) has not yet satisfactorily disposed of this point in terms of his theory. Also, the fact that zeatin, or other substances, occur in, or are isolated from, a given tissue source does not necessarily implicate them as an active growth substance. for they may be compartmented or complexed, so that the cells, in order to grow, may still need exogenous stimulation. For example, zeatin (its ribotide and riboside), has been isolated from particular cultures of soybean, which had become autonomous and required no exogenous cytokinin (although inositol was furnished). By contrast, other strains had an absolute requirement for such cytokinins (satisfied by zeatin); these are assumed *not* to produce the zeatin endogenously. Unfortunately, this was not tested, and therefore no strict conclusion can be drawn (Miura and Miller, 1969). At best, the zeatin is only one element in a growth-promoting system which, in turn, must integrate with a larger growth-regulating complex.

The cell division factor from crown gall tumor cells of *Catharanthus roseus (Vinca rosea)*, isolated by Wood and Braun (Wood *et al.*, 1969), is a purine derivative which is not N^6-substituted, for it has an oxygen in the 6-position. Although at first thought to be a substituted pyridinium compound related to nicotinamide (see Wood and Braun, 1967), the ideas about this compound have now changed and the suggested structure is 3,7-alkyl-2-alkylthio-6-purinone attached to a glucose moiety (Wood, 1970). Since this compound is not an N^6-substituted adenine it does not qualify as a cytokinin on the rigid definition of Skoog, but how then should this very active cell division factor be classified?

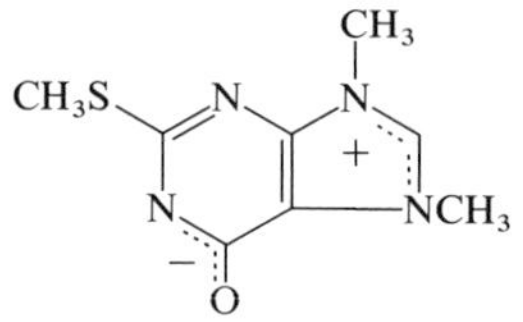

3,7-Alkyl-2-alkylthio-6-purinone

A wide range of compounds related to adenine have been synthesized and tested in a number of bioassays (Shaw *et al.*, 1968; Leonard *et al.*, 1968, 1969; Letham, 1968; Letham *et al.*, 1969; Rothwell and Wright, 1967; Skoog and Leonard, 1968). Certain structure/activity relationships have been shown to determine the chemical features necessary for adenyl cytokinin activity (Miller, 1970; Skoog and Armstrong, 1970; Skoog *et al.*, 1967; Letham, 1967b). Blockage of the number 1 position by the addition of a substituent seems to destroy activity as a cytokinin. The same may be true of the number 3 position. Substitution of the number 8 carbon and the number 9 nitrogen also destroys activity as a cytokinin. The length of the side chain also is important, e.g., 6-methylaminopurine is active although longer chains up to an optimum length of five carbons are more effective. It seems that unsaturation of the side chain (not its branching) is responsible for the high activity of zeatin and related compounds. Substitution of the nitrogen in the 6 position with oxygen or sulfur diminishes activity to 10% of its former activity. Optically active adenyl cytokinins have also been synthesized and assayed. Enantiomorphs of S configuration were shown to be more active than those of R configuration (Koshimizu *et al.*, 1968c).

Terpenes and Terpenoids

Gibberellic Acid and Related Compounds

Many gibberellins have been isolated from both fungal cultures and higher plants (see Brian, 1966; Lang, 1970; Tamura *et al.*, 1968). These are now regarded by many as a distinctive class of natural plant growth substances (MacMillan and Takahashi, 1968) since they have been isolated from and identified in several higher plant species, e.g., *Phyllostachys* (a bamboo), *Citrus, Phaseolus, Canavalia, Lupinus,* and *Calonyction* (the moonflower) (Brian *et al.,* 1967; Koshimizu *et al.,* 1968a; Tamura *et al.,* 1968). The number of plants from which gibberellins have been isolated is, however, still small, and they represent only a fraction of the plant kingdom (Lang, 1970). (Whether the legumes produce gibberellins with equal ease if grown nonsymbiotically is a moot point, as indeed is the possible fungal contact of other species from which gibberellins are isolated.)

There are 34 chemically identified gibberellins (Crozier *et al.,* 1970; Lang, 1970), and the list still increases. (See formulas of some of those added since May, 1969.) The designation of A_1, A_2, A_3, and A_4 were originally used by the Japanese workers to describe the first four gibberellins. The trivial names were continued by later workers to describe gibberellins A_5 to A_{17}. Trivial names for the gibberellins from bamboo, *Pharbitis, Canavalia,* and *Lupinus* were fortuitously withheld until agreement was reached to extend the gibberellin A_1 . . . A_n system to include the higher plant gibberellins. The current system involves allocation of gibberellin "A numbers" to "naturally occurring, fully characterized compounds which possess the gibbane skeleton and the appropriate biological properties" (MacMillan and Takahashi, 1968). The gibberellin A numbers do not necessarily follow the order of discovery. Gibberellic acid, for instance, was the first characterized gibberellin, even though it is designated gibberellin A_3.

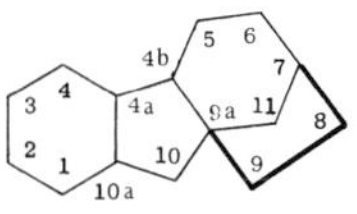

Gibbane

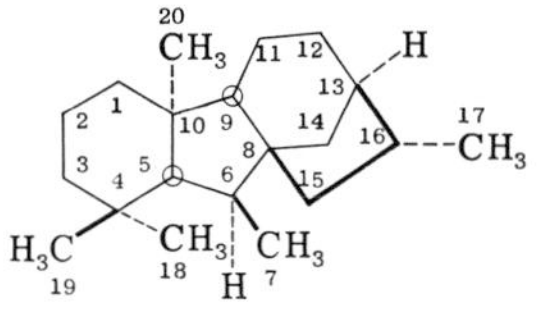

ent-Gibberellane

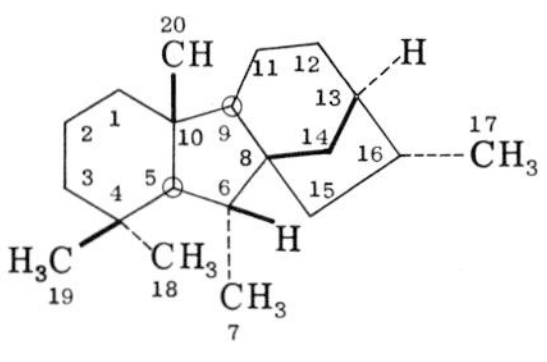

Gibberellane

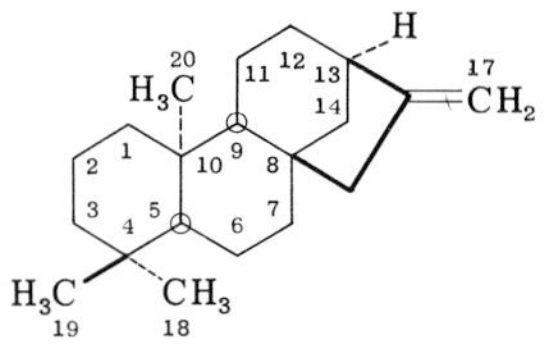

ent-Kaurene
[(−)-kaurene]

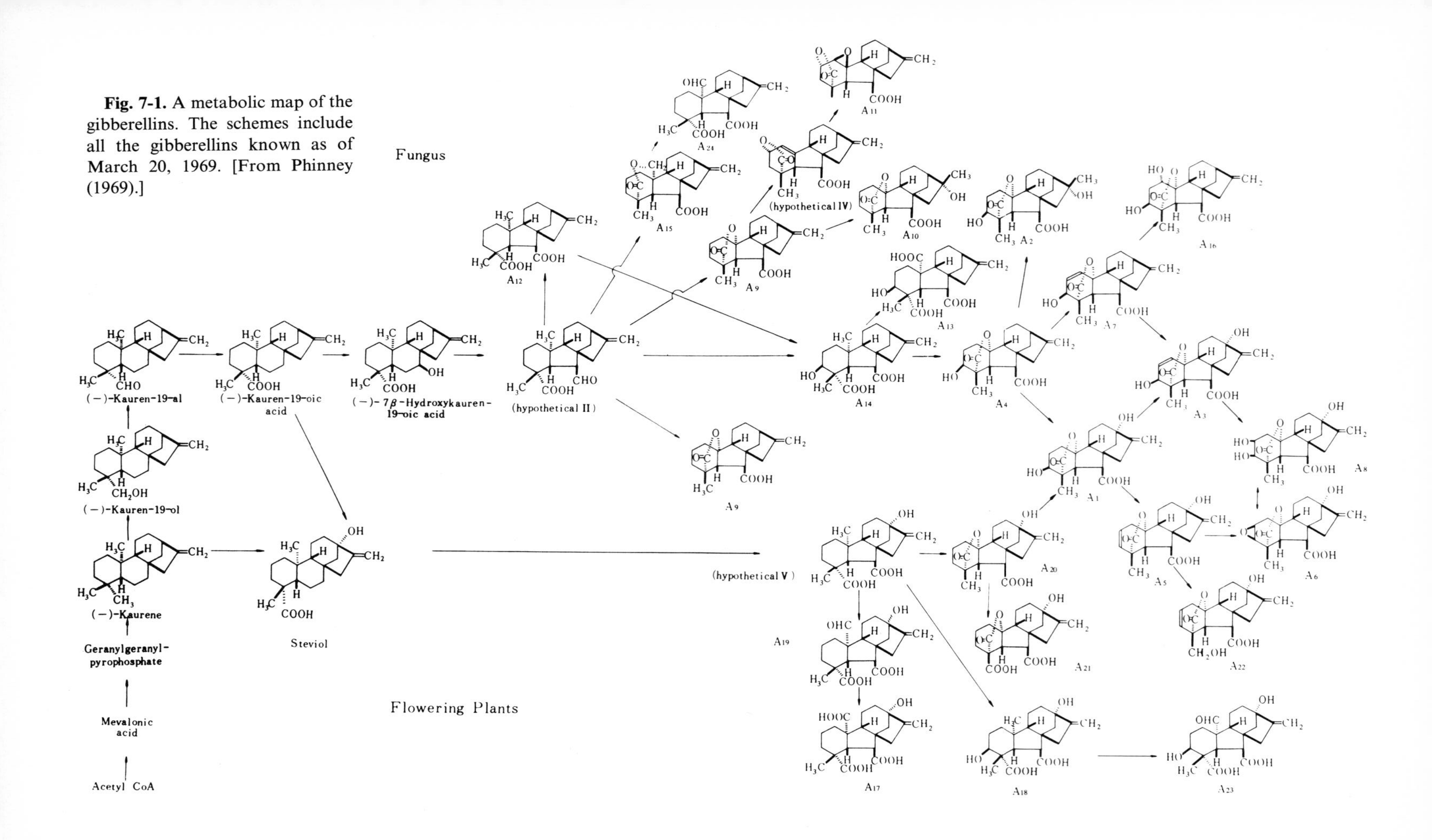

Fig. 7-1. A metabolic map of the gibberellins. The schemes include all the gibberellins known as of March 20, 1969. [From Phinney (1969).]

The specific nomenclature of the gibberellins has been based upon the gibbane nucleus, but a new convention suggests that another parent compound "gibberellane" be adopted. The gibberellane skeleton is sterically defined and has a numbering system corresponding to those of other cyclic diterpenes, such as kaurene. Thus the gibberellins would all be based on the enantiomer of gibberellane (*ent*-gibberellane) and the (−)-kaurene intermediary would be designated *ent*-kaurene. The only point of inconvenience in the system is that the corresponding enantiomer gibberellane is not representative of any naturally occurring compounds (see Lang, 1970 and references cited therein). Conjugated gibberellins, e.g., gibberellin glucosides (see Lang, 1970), are also known. Hitherto, all the gibberellins have been regarded as derived from fluorene-9-carboxylic acid, which, in turn, is derived from *ent*-kaurene (Cross, 1968; West and Upper, 1969). However, other naturally occurring compounds, helminthosporal, helminthosporic acid, and dihydrohelminthosporic acid, isolated from the plant pathogen *Helminthosporium sativum,* have been shown to elicit certain physiological responses similar to some (but not all) of those of gibberellic acid (Taniguchi and White, 1967). Other examples are sclerin and sclerolide (Tokoroyama *et al.,* 1969) from the fungus *Sclerotinia libertiana,* which also show gibberellin-like activity (Satomura and Sato, 1965) but are chemically different. Thus, views on the type of molecule needed to yield gibberellin-like activity must be expanded. Structure/activity relationships of a large number of natural and synthetic compounds have been determined (Bachi *et al.,* 1969; Brian *et al.,* 1967; Brinks *et al.,* 1969; Crozier *et al.,* 1970), but no generalizations from the studies of structures are possible. As far as the structure is concerned, the main difference seems to be whether there are 19 or 20 carbon atoms and whether hydroxyl (OH) groups occur at 3 and 13. Those gibberellins with 20 carbons (e.g., gibberellins A_{12}–A_{15}, A_{17}–A_{19}, A_{23}–A_{25}, A_{27}, A_{28}) have carboxyls in positions 7 and 18 and some (e.g., gibberellins A_{13}, A_{17}, A_{25}, A_{28}) also in position 20, whereas others (gibberellins A_{19}, A_{23}, A_{24}) have an aldehyde at position 20. Those gibberellins with 19 carbon atoms (A_1–A_{11}, A_{16}, A_{20}–A_{22}, A_{26}, A_{29}) all have a carboxyl at position 7 and are lactonic in the so-called ring "A." This causes the carbon difference between the C_{19} and C_{20} gibberellins.

It has been pointed out that the presence or absence of hydroxyls in positions 3 and 13 seems to distinguish those gibberellins that occur only in the fungus *Gibberella fujikuroi* (*Fusarium moniliforme*) and those that occur both in the fungus and in higher plants, or only in higher plants (see Lang, 1970). If the fungal gibberellins only possess one hydroxyl, it is invariably in the 3 position, whereas the higher plant gibberellins obtained thus far always have the hydroxyl in position 13. These differences have been interpreted as reflecting two major biosynthetic pathways, one of which is common to the fungus and higher plants, the other existing only in higher plants. Phinney (1969) has produced a chart (Fig. 7-1) of the numerous gibberellins known prior to March 20, 1969. This chart emphasizes the structural relationships and suggests that there are distinctive substances in the fungus and in flowering plants, respectively. However, no physiological roles for these substances are known in the fungus which produces them.

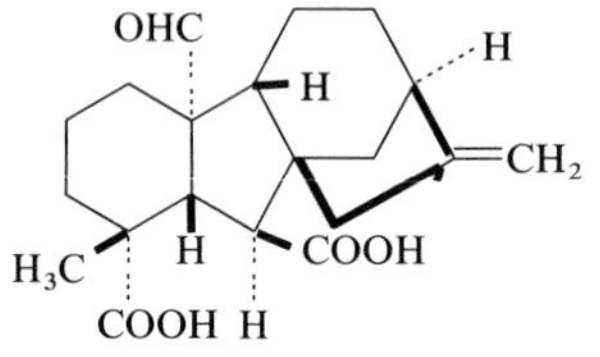

OHC H H CH_2 H_3C H COOH COOH H

Gibberellin A_{24}

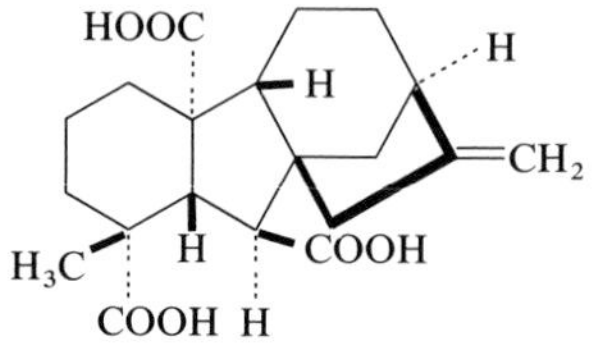

HOOC H H CH_2 H_3C H COOH COOH H

Gibberellin A_{25}

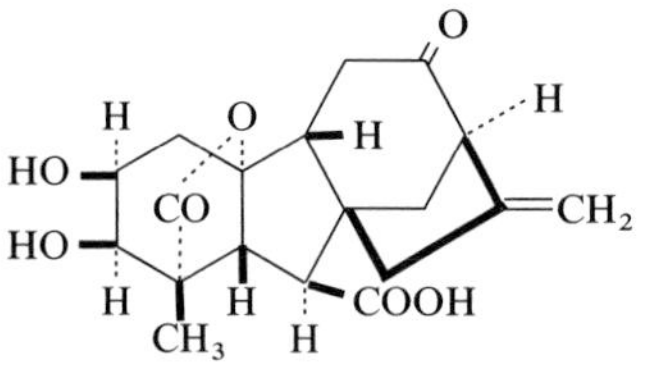

O H O H H HO HO CO CH_2 H H COOH CH_3 H

Gibberellin A_{26}

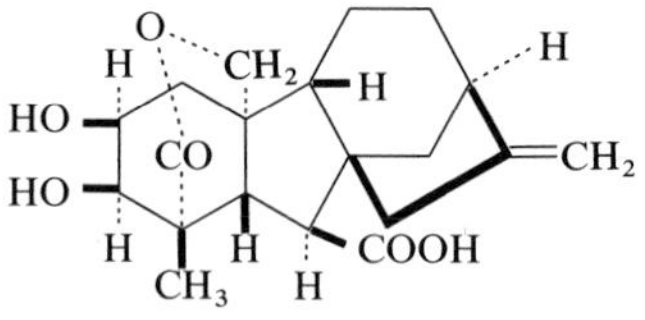

O CH_2 H H H HO HO CO CH_2 H H COOH CH_3 H

Gibberellin A_{27}

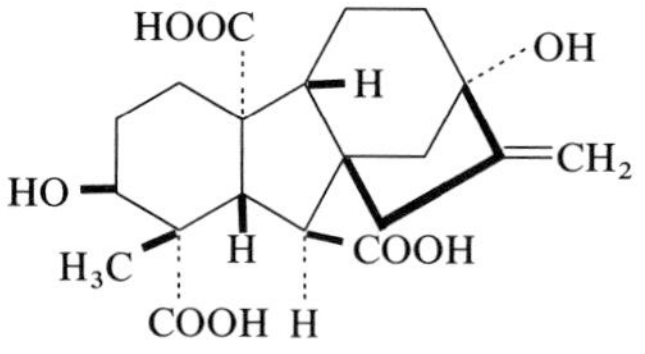

HOOC H OH CH_2 HO H_3C H COOH COOH H

Gibberellin A_{28}

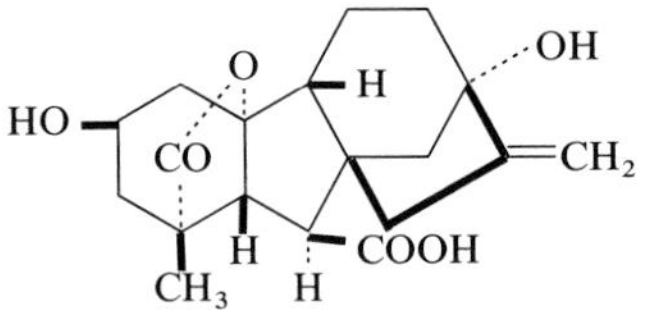

O H OH HO CO CH_2 H COOH CH_3 H

Gibberellin A_{29}

CH_2OH O HO HO OH O O H CO OH HO H CH_2 CH_3 COOH

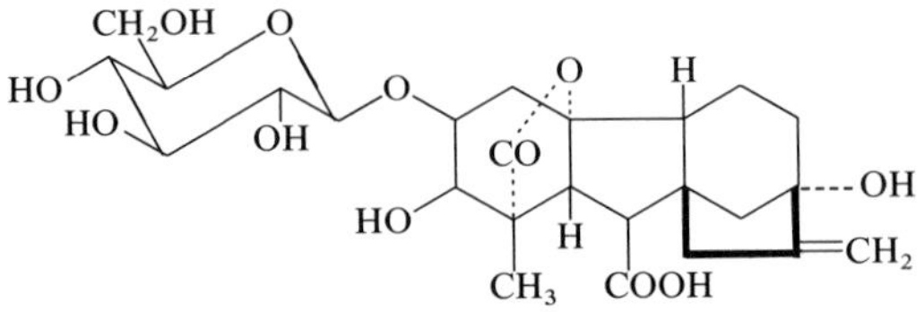

Phaseolus gibberellin
[*O*-(3)-*β*-D-Glucopyranosylgibberellin A_8]

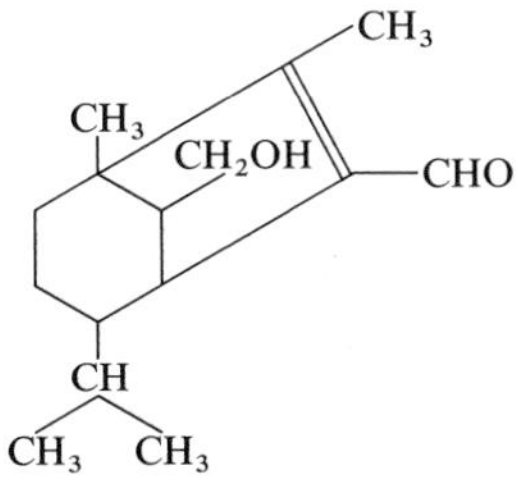

Helminthosporal

The "morphactins" are synthetic derivatives of fluorene-9-carboxylic acid (Schneider, 1970). Flurenol has been adopted as the common name for 9-hydroxyfluorene-(9)-carboxylic acid and the 2-chloro derivative is designated chlorflurenol. Some of the morphogenetic effects attributable to these substances are retardation of growth and interference with phototropism of shoots and with geotropism of roots (Khan, 1967; Krelle and Libbert, 1967; Mohr and Ziegler, 1969; Schneider, 1970; Ziegler, 1970).

The morphactins are absorbed through the leaves and roots and are translocated acropetally and basipetally. There is a general inhibition of elongation of internodes, a reduction of laminar area, an inhibitory effect on apical dominance which results in deformed growth of axillary shoots, and a stimulation of axillary bud growth. They inhibit elongation of the main axis and promote branching thus yielding a compact, bushy-shaped plant. Effects on the root system are characterized by a strong inhibition of lateral root formation (see Fig. 7-2A). When applied at appropriate stages of growth, there is an inhibition of flower formation, or delay in flowering, or induction of parthenocarpic fruit development.

The morphactins do not influence organs or organ primordia which are already formed. These substances are thus conceived as operating on either apical cells or meristems by reducing the rate of cell division. Morphactins affect organs which develop after their application. Normal growth resumes fairly quickly after the affects of a single treatment. Therefore the course of action is slow but long lasting with continuous or regular treatment. The action of morphactins, in short, is one that is inhibitory to development, but the spectrum of response is very broad (see Mohr and Ziegler, 1969; Schneider, 1970; Ziegler, 1970; and references cited therein).

Horticulturally and agriculturally, morphactins at the appropriate dosage are being tested as agents that inhibit flower bud formation, that prevent flowering, or that cause premature shedding of flowers. For instance, they delay or prevent bolting in lettuce but promote "head" formation. Morphactins also act synergistically with auxins such as 2,4-D and substances such as maleic hydrazide (e.g., the susceptibility to retardant and herbicidal activity is enhanced by the morphactins).

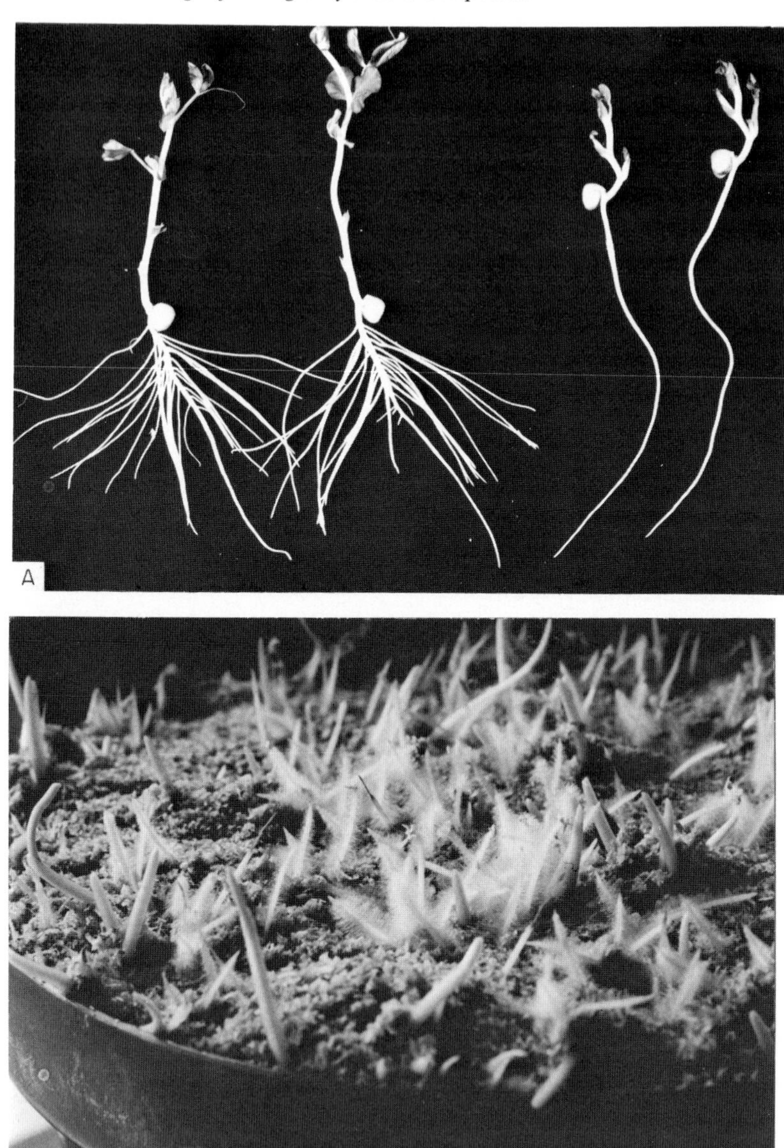

Fig. 7-2. A, Ten-day-old seedlings of *Pisum sativum*. Left, two control plants grown in water; right, two plants from seeds that had been soaked 12 hours in 5×10^{-5} *M* chlorflurenol. The morphactin has caused shortening of the internodes, inhibition of leaf blade growth, and a suppression of lateral roots. There is no effect on the elongation of the primary root.

B, *Avena sativa* (4 days old) whose grains had been soaked in 10^{-5} *M* chlorflurenol. The roots are growing upwards (negatively geotropic) while the coleoptiles (distinguished by lack of hairs) are growing normally. (Photographs courtesy of Dr. H. Ziegler, Technische Hochschule, München, West Germany.)

Morphactins were called "antigibberellins" without actual evidence of competitive inhibition (Krelle and Libbert, 1967). This view derived from the observation that the inhibitory effects on elongation by morphactins could be partially reversed by gibberellins. On closer examination (see Ziegler, 1970) it became clear that the three-dimensional structure of morphactins and gibberellins are really very different and there was no direct interaction between the two classes of growth-regulatory substances. For instance, morphactins have no effect on the α-amylase activity of barley endosperm in that bioassay system (see Schneider, 1970, and references cited therein).

A curious feature of morphactin activity is that their affects on pea shoots (inhibition of elongation, for example) is much enhanced by darkness. The opposite has been shown for the true antigibberellin, chlorocholine chloride (CCC) (see page 106) which also causes "dwarfing." The interpretation here is that growth in darkness depends on a system sensitive to chlorflurenol, and growth in light is sensitive to CCC. The dramatic negative geotropism (see Fig. 7-2B) which results from flurenol treatment is probably due to suppressed lateral transport of indole-3-acetic acid (IAA) (see Ziegler, 1970). Interestingly enough, roots growing upwards under the influence of morphactins show a normal displacement of statoliths in the root cap. This means that the direction of curvature of a root need not be determined, as is often supposed (see Audus, 1969), by the position of the statolith amyloplasts alone (see Ziegler, 1970).

It is also interesting to note that esters of benzilic acid (diphenylglycolic acid) also show activity identical with that of the prototype morphactins (see Dalton and Brown, 1970). Dalton and Brown (1970) have suggested that the degree of activity of substituted benzilic acid molecules could be related to the planarity of the two aromatic rings.

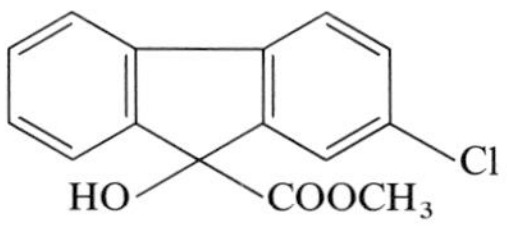

Cl
HO COOCH$_3$

Methyl-2-chloro-9-hydroxyfluorene-(9)-carboxylate

Gibberellic acid has been found to shorten instars in locust development, and ecdysone itself (20-deoxyecdysterone) causes a significant growth promotion of dwarf pea seedlings (Brian, 1966). There is now an active area of research which deals with insect development and gibberellins and it has even been supposed that ecdysones are "gibberellins" or closely related compounds. Ecdysterone and related steroid hormones which control molting in insects have been isolated from plants (Heftmann, 1970; Staal, 1967; Williams, 1969). Moreover, the juvenile hormone of insects is closely related to farnesol, a component of many essential oils (Stowe and Hudson, 1969). In this important area of plant–animal relations, however, speculation is ahead of knowledge.

"Abscisins" or "Dormins"

In 1963 a substance [3-methyl-5-(1-hydroxy-4-oxo-2,6,6-trimethyl-2-cyclohexene-1-yl)-*cis,trans*-2,4-pentadienoic acid] from young cotton bolls was isolated and named Abscisin II (see Addicott and Lyon, 1969).

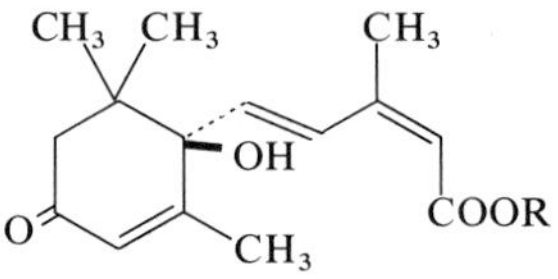

Abscisic acid,
R = H

(+)-Abscisyl-β-D-glucopyranoside,
R = glucose

The name was later changed to abscisic acid (ABA). A parallel investigation by Wareing and co-workers disclosed the presence of a substance in the leaves of deciduous trees which inhibited growth and induced dormancy (Wareing, 1969a, b; Wareing and Ryback, 1970). This substance was named "dormin," but it was soon found to be abscisic acid (El-Antably *et al.*, 1967). Abscisic acid is widely distributed in the plant kingdom and has a number of physiological effects besides inducing abscission and dormancy (Addicott and Lyon, 1969; Wareing and Ryback, 1970) (see Fig. 7-3).

The chemical structure was determined by Ohkuma *et al.* The stereochemistry of this highly asymmetric carbon, with an unusually high specific rotation, was established by the synthesis of Cornforth *et al.* (Addicott and Lyon, 1969). The structure is remarkably similar to that of the carotenoids, especially violaxanthin, and it has been suggested that abscisic acid actually derives from carotenoids (Taylor and Smith, 1967; Taylor, 1968). Since the function of carotenoid pigments in plants is only poorly understood, the finding that certain xanthophylls may act as precursors to abscisic acid is attractive, especially since it implicates some of the responses of plants to blue light. Although much recent emphasis has been placed on the xanthophyll origin of abscisic acid, the alternative mevalonic acid pathway likewise leading to the formation of abscisic acid should not be overlooked (see Milborrow, 1970). In fact, the carotenoid pathway stresses light, whereas the mevalonic acid pathway may operate more in darkness in a manner analogous to animal (liver) systems. Various analogs have

Fig. 7-3. Effects of abscisic acid and length of day on carnation plants. In each group the first plant was not treated with the regulator, the second received 500 ppm, and the third received 1000 ppm. Left, plants grown with 8 hours light per day; center, plants grown with 12 hours light per day; right, plants grown with 24 hours light per day. Note that the abscisic acid retarded elongation of internodes in the axis, and it suppressed the floral stimulus otherwise due to long days. (Photograph courtesy of the U. S. Department of Agriculture.)

been synthesized and tested for activity (Addicott and Lyon, 1969; Okamoto *et al.*, 1970; Oritani and Yamashita, 1970; Sondheimer *et al.*, 1969; Tamura and Nagao, 1969a, b). Methyl esters appear to have the same activity as the parent compound, but other analogs are considerably less active than abscisic acid (Mousseron-Canet *et al.*, 1970). Abscisic acid also occurs as its glycoside (Koshimizu *et al.*, 1968b). Phaseic acid, an analog of abscisic acid isolated from *Phaseolus multiflorus,* shows weak activity in the cotton abscission bioassay (MacMillan and Pryce, 1968). Examination of the structural formulas of abscisic acid and phaseic acid will show that ABA can undergo rearrangement to yield phaseic acid. The role of phaseic acid, however, remains obscure.

Another member of the "ABA-family" of compounds has been discovered and named xanthoxin. This neutral compound is as active as ABA and can cause growth retardation, i.e., in height, but has the property of maintaining good lush, dark green leaves, etc. The implication here is that one may have a class of substances that are able to inhibit growth in the light but do not cause any visible deterioration of the system. Presumably, a long exposure to darkness would release any inhibition due to accumulation of these compounds. These findings add a new dimension to the effects of light on plant growth regulation (see Taylor and Burden, 1970a, b).

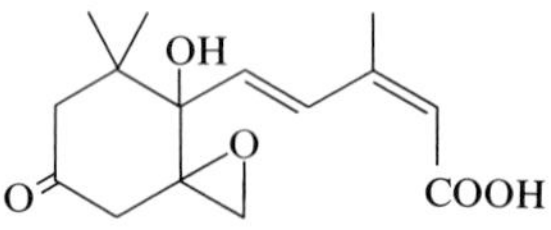

Phaseic acid

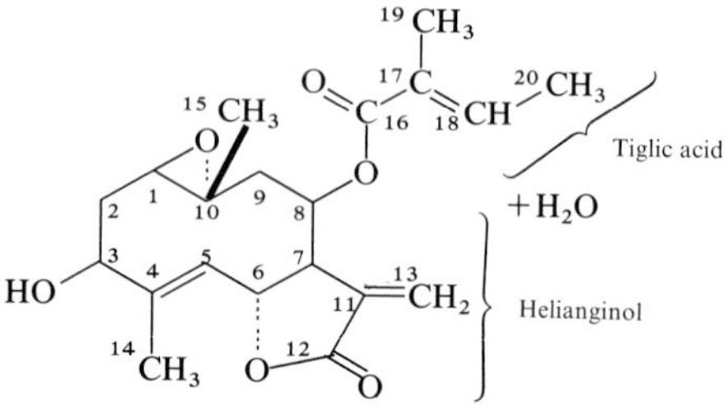

Heliangine

Heliangine, a sesquiterpene lactone (Nishikawa *et al.*, 1966) isolated from the leaves of sunflower, *Helianthus tuberosus*, is a regulator found to have inhibiting effects on *Avena* curvature and straight growth tests, but promotes adventitious root formation in *Phaseolus* cuttings. Moreover, heliangine is supposed to be the agent by which high light intensity suppresses elongation of stem internodes, for light accelerates heliangine transport from sunflower leaves and this, in turn, inhibits stem elongation in sunflower (Morimoto *et al.*, 1966).

Work on animal tumor inhibitors obtained from plants has disclosed several sesquiterpene dilactones [elephantin and elephantopin from *Elephantopus elatus* (Kupchan *et al.*, 1966) and vernolepin from *Vernonia hymenolepsis*] to be strong inhibitors of elongation in the wheat coleoptile section test (Sequeira *et al.*, 1968). If the inhibitors are washed out and then treated with an auxin, the segments then elongate. The reversibility of the action of these compounds has suggested that such compounds may play a regulating role *in situ* (Kupchan, 1970). The similarity of the structure of elephantin and elephantopin with that of heliangine is striking.

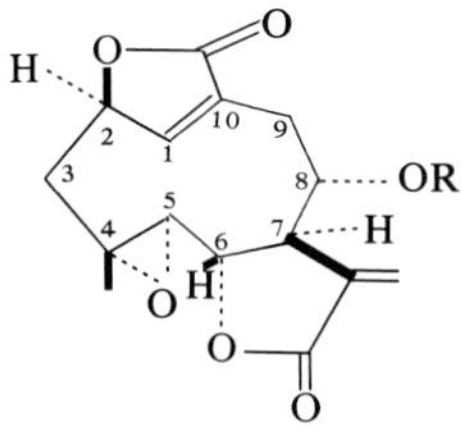

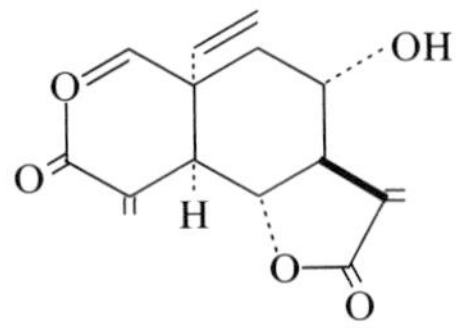

Vernolepin

Elephantin, R: $(CH_3)_2C=CHCO-$
Elephantopin, R: $CH_2=C(CH_3)CO-$

Pyrethrosin and "cyclopyrethrosin acetate" (its cyclization product) are sesquiterpene lactones isolated from *Chrysanthemum cinerariaefolium*, the insecticidal "pyrethrum daisy" (Iriuchijima and Tamura, 1970). The structural similarities for all these compounds suggest they may have similar biological activity (Nishikawa *et al.*, 1966).

Xanthinin, another terpenoid, has been isolated from *Xanthium pennsylvanicum*, the cocklebur. It has growth inhibitory effects. However, since the amount found in leaves is so high (1%), there is reason to doubt as to whether it acts as a growth regulator *in situ* or is highly compartmentalized (Deuel and Geissman, 1957; Geissman *et al.*, 1954).

Other Terpenes

Evenari (1949) lists a large number of compounds that act as growth and germination inhibitors. Essential oils, rich in terpenes, have been known for many years and used as germicidal agents, e.g., thymol (Garb, 1961; Toole *et al.*, 1956). There has been a great deal of work dealing with the allelopathic control of plant growth by terpenoid-containing species (Evenari, 1961; Grümmer, 1961; Muller, 1966, 1970; Rice, 1967).

Lactones

Lactones such as coumarin, scopoletin, parasorbic acid, and daphnetin can be found in higher plants (Berrie *et al.*, 1968; Evenari, 1949; Moreland *et al.*, 1966; Toole *et al.*, 1956). They may perform a regulatory role (Despois, 1958; van Sumere and Massart, 1958) and there is a great deal of information dealing with these and similar compounds as germination inhibitors and growth inhibitors (especially of roots). Some, e.g., scopoletin, may be growth-promoting at appropriate concentrations.

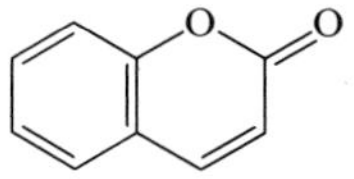

Coumarin (1,2-benzopyrone)

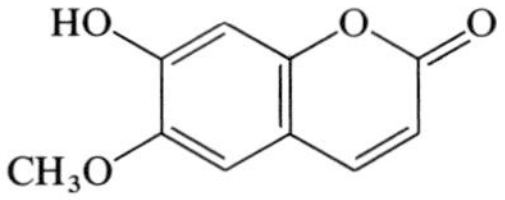

HO, CH_3O

Scopoletin (7-hydroxy-6-methoxycoumarin)
Scopolin (mono-β-glucopyranoside of scopoletin)

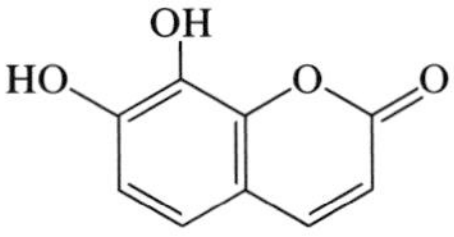

OH, HO

Daphnetin (7,8-dihydroxycoumarin)

Carbamates and Their Derivatives

Carbamates have been known for years to retard growth selectively (see Fig. 7-4; Ivens and Blackman, 1949). On exposure of plants to some

Fig. 7-4 (A)

of these, e.g., *O*-isopropyl-*N*-phenylcarbamate or isopropyl-*N*-(3-chlorophenyl)carbamate, mitotic activity ceases immediately (see Audus, 1964). Interference with photosynthetic activity has also been shown. AMO 1618 (4-hydroxy-5-isopropyl-2-methylphenyltrimethylammonium chloride 1-piperidine carboxylate) is a quarternary ammonium carbamate which is

Fig. 7-4 (B)

Fig. 7-4. The inhibition of sprouting of potato tubers. A, Untreated tubers in storage; B, tubers after dipping in a solution of chloroisopropylphenylcarbamate. (Photograph courtesy of the U. S. Department of Agriculture.)

widely used as a growth retardant (Baldev *et al.*, 1965; Cathey, 1964; West and Upper, 1969).

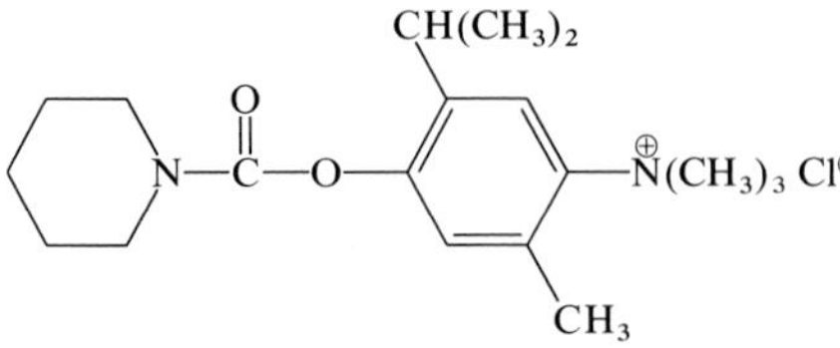

$$\text{C}_5\text{H}_{10}\text{N}-\overset{\text{O}}{\overset{\|}{\text{C}}}-\text{O}-\text{C}_6\text{H}_2(\text{CH}(\text{CH}_3)_2)(\text{CH}_3)-\overset{\oplus}{\text{N}}(\text{CH}_3)_3\ \text{Cl}^{\ominus}$$

AMO 1618

4-Hydroxy-5-isopropyl-2-methylphenyltrimethylammonium chloride 1-piperidine carboxylate

Carboxymethyl dimethyl dithiocarbamate and certain of its analogs can be shown to have high growth-promoting activity (Audus, 1964; Mann *et al.*, 1967).

Substituted Ureas

N′-(4-Chlorophenyl)-*N,N*-dimethylurea (monuron), *N′*-(3,4-dichlorophenyl)-*N,N*-dimethylurea (diuron), *N,N*-dimethyl-*N′*-phenylurea (fenuron), *N*-butyl-*N′*-(3,4-dichlorophenyl)-*N*-methylurea (neburon) are only a few of many urea derivatives that are used as herbicides. They exert their effects by interfering with the Hill reaction of photosynthesis (see Audus, 1964; Barth and Mitchell, 1969; Izawa and Good, 1965; Moreland, 1967).

A number of urea derivatives may also act as cell division factors. The first of these to be so described was *N,N′*-diphenylurea, isolated during fractionation of coconut milk by Shantz and Steward (1955). Although initially the synthetic compound showed sporadic activity, this is now known to be due to the impurity of some samples tested (Kefford *et al.*, 1968). A very large number of urea derivatives have now been synthesized and tested but, to date, it is not possible to state rigid structure/activity relationships for these compounds (Bruce and Zwar, 1966; Kefford *et al.*, 1966, 1968).

Triazine Compounds

In the early 1950's, triazines were introduced as herbicidal chemicals (see Audus, 1964). 2-Chloro-4,6-bis(ethylamino)-1,3,5-triazine (simazine) and 6-isopropylamino-1,3,5-triazine (atrazine) are the ones most widely used. The triazines also inhibit the Hill reaction of photosynthesis (Izawa and Good, 1956; Moreland, 1967).

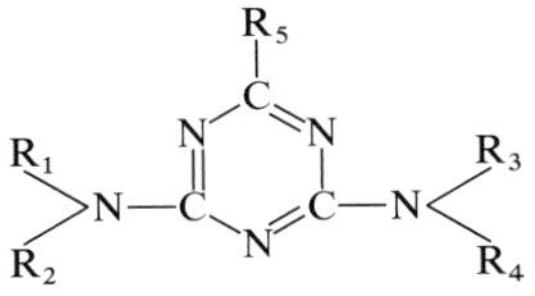

$$\begin{array}{c} R_5 \\ | \\ \text{triazine ring: } (R_1)(R_2)N\text{—C} \;\; C\text{—}N(R_3)(R_4) \end{array}$$

Triazines

R_1 and R_3=H or ethyl
R_2 and R_4=alkyl
R_5=Cl, —OCH_3, —SCH_3

Hydrazine Derivatives

Maleic hydrazide (Evenari, 1949; Greulach and Plyler, 1966; Moreland *et al.*, 1966) (the diethanolamine salt of 1,2-dihydroxy-3,6-pyridazinedione, B-9 (*N*-dimethylaminosuccinamic acid (Cathey, 1964), and *β*-hydroxyethylhydrazine (Palmer *et al.*, 1967) are derivatives of hydrazine that have found active use as growth regulators. The first is an inhibitor, the second a retardant, and the third promotes flowering of pineapples.

Attempts to evaluate other hydrazine derivatives have disclosed other substances, e.g., *N*-amino-*N*-methyl-*β*-alanine, which are growth inhibitors (Greulach and Plyler, 1966; Huffman *et al.*, 1968).

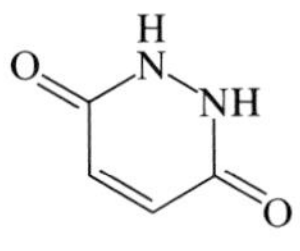

Maleic hydrazide
(1,2-dihydro-3,6-pyridazinedione)

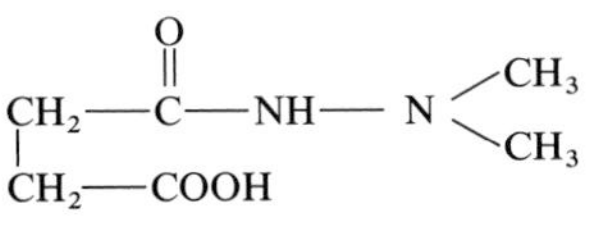

$$\begin{array}{l} CH_2\text{—}\overset{O}{\overset{\|}{C}}\text{—NH—N}(CH_3)_2 \\ | \\ CH_2\text{—COOH} \end{array}$$

N-Dimethylaminosuccinamic acid (B-9; Alar)

$$HO\text{—}CH_2\text{—}CH_2\text{—NH—}NH_2$$

β-Hydroxyethylhydrazine

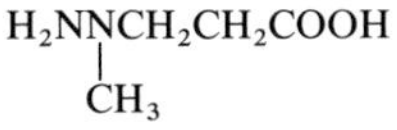

$$\begin{array}{l} H_2N\underset{|}{N}CH_2CH_2COOH \\ \quad\;\; CH_3 \end{array}$$

N-Amino-*N*-methyl-*β*-alanine

Quarternary Ammonium Compounds

In 1960, a new group of quarternary ammonium growth regulators was recognized (Tolbert, 1960). (2-Chloroethyl)trimethylammonium chloride was the most active, and it is now extensively used as a retardant (Cathey, 1964). Its short designation (CCC) derives from chlorocholine chloride.

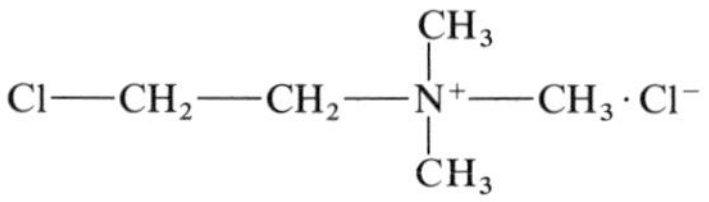

$$Cl—CH_2—CH_2—\overset{\overset{CH_3}{|}}{\underset{\underset{CH_3}{|}}{N^+}}—CH_3 \cdot Cl^-$$

(2-Chloroethyl)trimethylammonium chloride (CCC)

AMO 1618 (4-hydroxy-5-isopropyl-2-methylphenyltrimethylammonium chloride 1-piperidine carboxylate) has already been mentioned as a quarternary ammonium carbamate (Cathey, 1964). Many quarternary ammonium compounds such as morpholinium, picolinium, and quinolinium have also been tested (Cathey, 1964).

Phosphoniums

2,4-Dichlorobenzyltributylphosphonium chloride (Phosphon D) is another popular retardant that has been recently developed (Cathey, 1964) (Fig. 7-5). Phosphoniums have been recognized as inhibitors for a long time (see Audus, 1964).

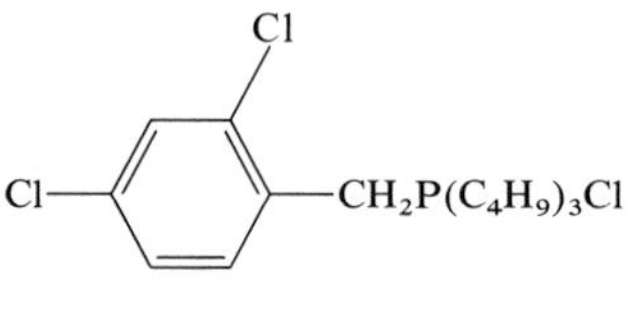

Cl (ring, 2-position); Cl— (ring, 4-position); $—CH_2P(C_4H_9)_3Cl$

2,4-Dichlorobenzyltributylphosphonium chloride (Phosphon D)

Phenolics

Many naturally occurring phenolic compounds are known germination inhibitors (Despois, 1958; Evenari, 1949; Mayer and Poljakoff-Mayber,

Fig. 7-5. The interaction of Phosphon with environmental factors in the growth and flowering of *Rhododendron*. (The plants were grown for 4 months on long days, and on short days for 2 months; they were chilled at 50°F for 2 months.) Left, plant not treated with the growth regulator; right, plant treated with Phosphon. (Photograph courtesy of the U. S. Department of Agriculture.)

1963; Moreland *et al.*, 1966; Nikolaeva, 1969; Toole *et al.*, 1956; Wang *et al.*, 1967). Phloridzin (2′,4′,6′-4-tetraoxydihydrochalcone-2′-glucoside) the glucoside of phloretin which occurs in apple (*Malus*) has long been

HO—C_6H_4—CH_2—CH_2—C(=O)—C_6H_2(O—glucose)(HO)(HO)

Phloridzin

known to be a potent inhibitor of growth (Hutchinson *et al.*, 1959; Kefeli *et al.*, 1969). Moreover, some, e.g., chlorogenic acid and juglone (Moreland *et al.*, 1966; Sondheimer, 1964) inhibit growth and may be important allelopathic compounds (Börner, 1960; Evenari, 1961; Muller, 1966; Rice,

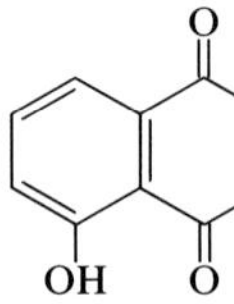

Juglone
(5-hydroxy-1,4-naphthoquinone)

1965, 1967), but at lower concentrations they may even be stimulants of cell division. The known relations of phenolic substances (mono- and di-) to oxidations in plants have often been implicated in the control of the stability of IAA *in situ* (Hanson *et al.*, 1967; Ray, 1958). Cinnamic, caffeic, coumaric, ferulic acids, etc., also are known inhibitors (Börner, 1960; Moreland *et al.*, 1966; Sondheimer, 1964). The number of identified inhibitory phenolic compounds is growing, as work aimed at the metabolism of a phenolic inhibitor from *Salix* (willow), identified as chalconaringenin-2′-glucoside, shows (Turetskaya *et al.*, 1968; Kefeli *et al.*, 1969).

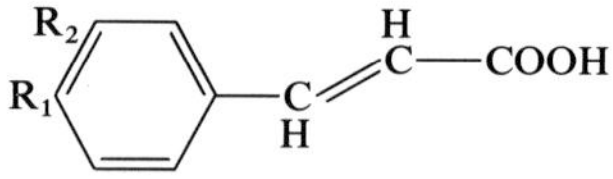

trans-Cinnamic acid, $R_1 = R_2 = H$
trans-p-Coumaric acid, $R_1 = OH$, $R_2 = H$
trans-Caffeic acid, $R_1 = R_2 = OH$
trans-Ferulic acid, $R_1 = OH$, $R_2 = OMe$

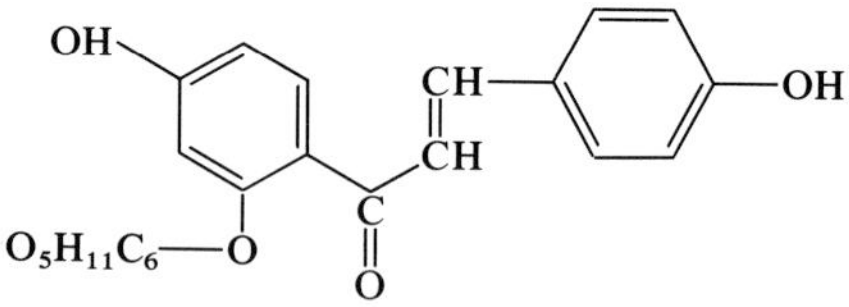

Chalconaringenin-2′-glucoside

Miller (1969) has shown that cultured soybean tissue (of cotyledonary origin) synthesizes large amounts of daidzin (4′,7-dihydroxyisoflavone glucoside) and related compounds in response to the addition of cytokinins in an otherwise deficient medium. (This occurs only in the presence of auxin however.) Because daidzein (the aglucone of daidzin), formonetin (7-hydroxy-4′-methoxyisoflavone), and genistin (the glucoside of 4′,5,7-trihydroxyisoflavone) showed no effects as auxins or cytokinins in a number

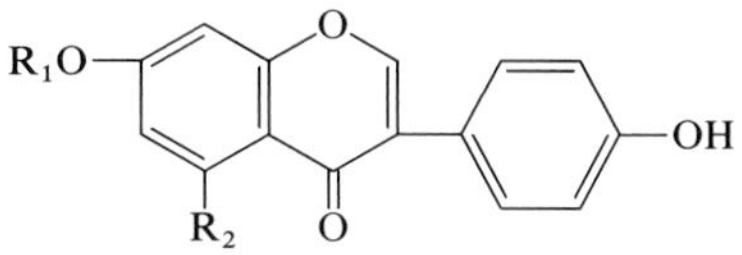

Daidzein: R_1 and R_2=H
Daidzin: R_1=glucose, R_2=H
Genistin: R_1=glucose, R_2=OH

of assays it seems unlikely that the production of deoxyisoflavones is the means by which the growth regulatory substances (such as the adenyl cytokinins) act (Miller, 1969). Nevertheless, Miller (1969) suggests that the isoflavone *precursors* might play an important role in growth and differentiation. It is known however that cell division is not involved in the response to the stimulus to synthesize the deoxyisoflavones since the production of similar phenolic compounds is stimulated by adenyl cytokinins and diminished by inhibitors such as chloramphenicol, cycloheximide, and actinomycin D, even when the cells are plasmolyzed (see Miller, 1970).

Other naturally occurring phenolic substances may act synergistically with auxins as growth promoters in various bioassay systems (Steward, 1968; Nitsch and Nitsch, 1962; Wain and Taylor, 1965). The class of phenolics, such as the leucoanthocyanins, are noteworthy in this connection, for they are capable of inducing cell division in the carrot root phloem bioassay and occur prominently in many fruits in the vicinity of developing embryos (Steward and Shantz, 1956).

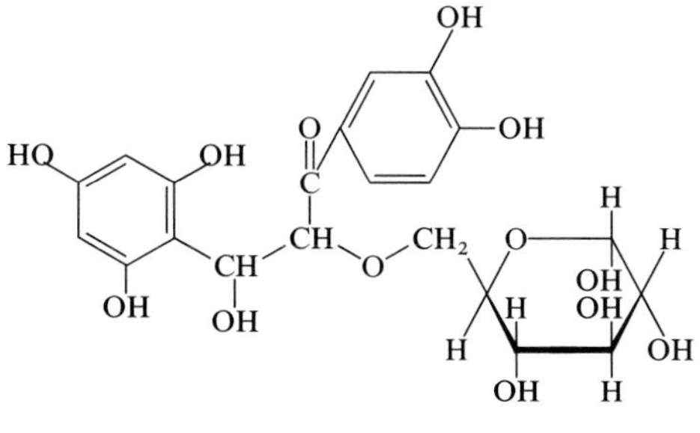

Leucocyanidin monoglucoside

Hexitols

Work on the biochemistry of the coconut milk growth factors now suggests that there are cell division and growth-promoting substances which require and interact with hexitols. Inositol represents the so-called neutral fraction of these naturally occurring, growth-promoting systems (Shantz and Steward, 1964). Pollard *et al.* (1961) isolated *myo*-inositol, *scyllo*-inositol, and D-sorbitol from natural sources such as coconut milk, corn in the milk stage, and *Aesculus* vesicular embryo sac fluids, but the role of hexitols in the carrot bioassay is adequately fulfilled by *myo*-inositol (Shantz and Steward, 1964, 1968). An auxin–inositol complex (indole-3-acetyl-2-*O*-*myo*-inositol) has also been isolated from *Zea mays* (Bandurski *et al.,* 1969; Nicholls, 1967).

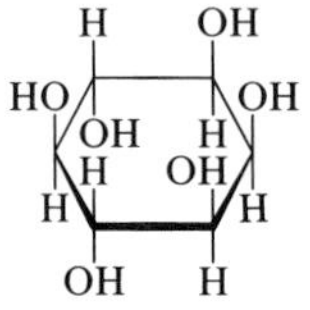

myo-Inositol

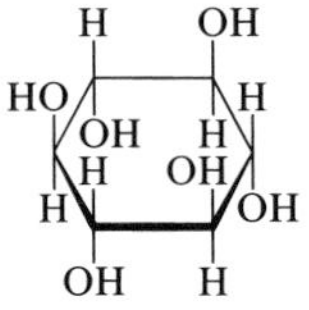

scyllo-Inositol

CH_2OH
HCOH
HOCH
HCOH
HCOH
CH_2OH

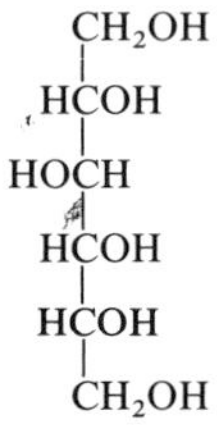

D-Sorbitol

Inositol and its phosphates (phytic acid) have long been known as conspicuous in seeds and fruits (Anderson and Wolter, 1966). From the early days of work on vitamins its claims have from time to time been advanced. It is known as a component part of phosphatides and in plants it enters into the structure of compounds (hexuronic acids) involved with the metabolism of pectins and the components of cell walls (Loewus, 1969). The current work on naturally occurring auxin–inositol complexes and on IAA–glycosides which interact *in vivo* with inositol opens up a broad area of involvement of the hexitols in the understanding of growth regulation (Bandurski *et al.,* 1969; Hoffman-Ostenhof, 1969; Loewus, 1969; Tanner, 1969).

Amino Acids and Peptides

A growth regulatory role has been suggested for some of the nonprotein amino acids that occur in the soluble nitrogen pools of plants. Azetidine-2-carboxylic acid is a proline antagonist, ethionine competes with methionine, and canavanine is an arginine antagonist (Fowden *et al.*, 1967, 1968). Other examples could be cited, but all these compounds, where active, have growth inhibitory effects—presumably because they replace at active sites the otherwise essential amino acid (Fowden *et al.*, 1967).

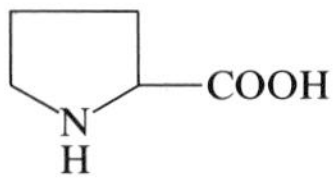

Proline

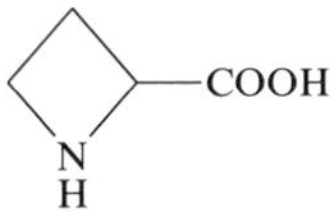

Azetidine-2-carboxylic acid

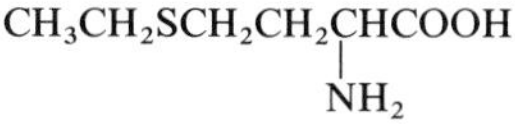

$CH_3CH_2SCH_2CH_2CH(NH_2)COOH$

Ethionine

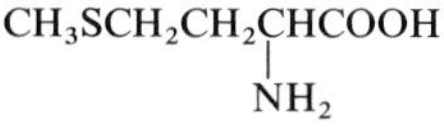

$CH_3SCH_2CH_2CH(NH_2)COOH$

Methionine

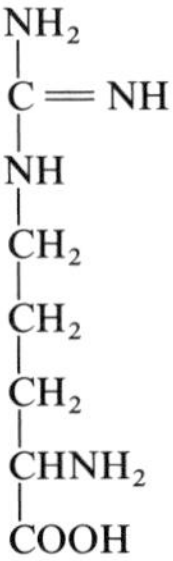

NH_2—C(=NH)—NH—CH_2—CH_2—CH_2—$CHNH_2$—COOH

Arginine

NH_2—C(=NH)—NH—O—CH_2—CH_2—$CHNH_2$—COOH

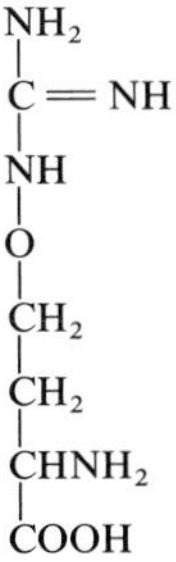

Canavanine

Besides many peptides which behave as "antibiotics" there are some that exert profound morphological effects (Brian, 1957). Malformin (comprised of valine, leucine, isoleucine, and half cystine), a product of *Aspergillus niger,* affects root development (Takahashi and Curtis, 1961).

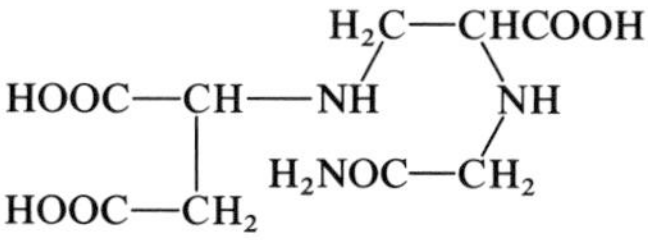

HOOC—CH—NH—H_2C—CHCOOH—NH—CH_2—H_2NOC; HOOC—CH_2

Lycomarasmin

Some other peptides that occur as toxins are products of fungal or bacterial activity but, nevertheless, they exert their effects as part of the disease syndrome, e.g., lycomarasmin which is implicated in vascular wilt diseases (Dimond, 1959; Owens, 1969).

No attempt therefore will be made to include here the products of microbial metabolism that have effects on plant growth. This would be a formidable task. Nevertheless, this is a very active research area and new compounds are being disclosed very rapidly. An example is fragin, a metabolite of

Pseudomonas fragi which contains a unique *N*-nitrosohydroxylamino group. Fragin is active as a growth inhibitor and has antitumor activity (Murayama *et al.*, 1969; Murayama and Tamura, 1970a,b).

Alkaloids

Although these nitrogenous bases are often dismissed as "secondary" plant products, they include many that show growth regulatory activity. Probably the best known of these is colchicine which interferes with mitosis (see Shelanski and Taylor, 1967, and references cited therein). Triacanthine

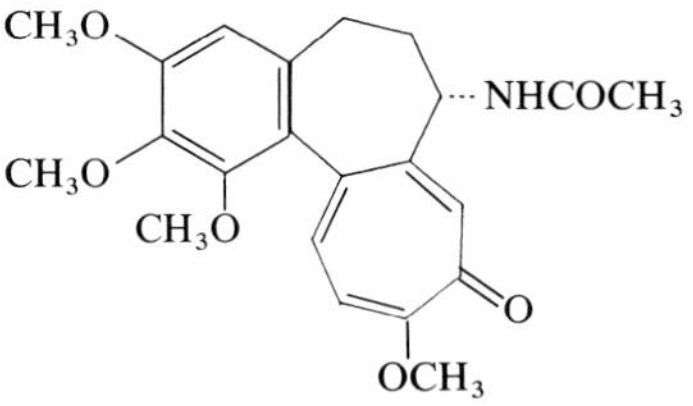

Colchicine

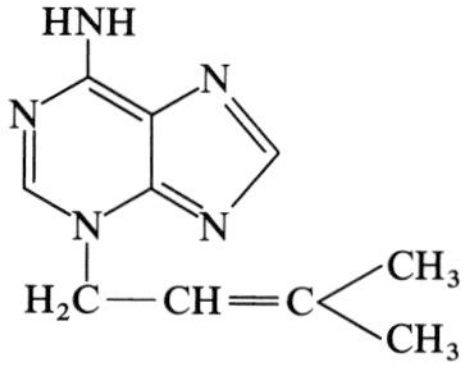

Triacanthine
[6-amino-3-(γ, γ-dimethylallyl)purine]

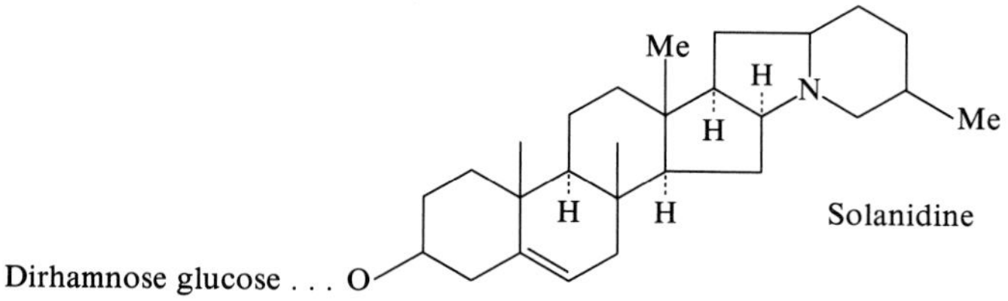

Solanidine

Dirhamnose glucose . . . O

α-Chaconine

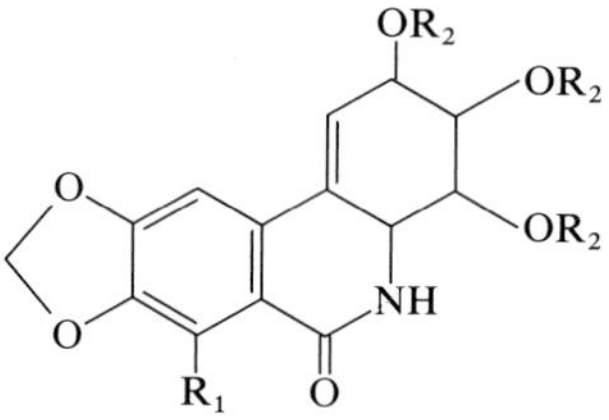

Lycoricidinol, $R_1 = OH$, $R_2 = H$
Lycoricidine, $R_1 = R_2 = H$

from *Gleditsia triacanthos*, the honey locust (Rogozinska *et al.*, 1964), has cytokinin activity when autoclaved and its structure is that of a N^3-adenyl "cytokinin" (Rogozinska, 1967). The view is that triacanthine is converted to a N^6-adenyl cytokinin during the heat sterilization process that often accompanies the preparation of tissue culture media (Skoog and Leonard, 1968). α-Chaconine (a solanidine glycoside) (Kuhn and Löw, 1954), isolated from the leaves of *Solanum* species, is a potent growth inhibitor (Steward, 1968), and so are lycoricidine and lycoricidinol from bulbs of *Lycoris radiata* (Okamoto and Torii, 1968). (Narciclasine from daffodil bulbs is identical with lycoricidine.)

Waller and Burström (1969) have tested the inhibitory activity of the diterpenoid alkaloid delcosine from *Delphinium ajacis*, the common larkspur. The similarity of structure between delcosine and gibberellic acid have suggested to these workers that delcosine may compete with gibberellic acid for enzyme-active sites and thereby change their catalytic function.

Thus, there is much information that implicates alkaloids both as growth promotors and growth and germination inhibitors (see Côme, 1970; Evenari, 1949; Moreland *et al.*, 1966; Toole *et al.*, 1956).

Glycosides

Some glycosides have already been mentioned, e.g., glucobrassicin (Kutáček, 1967), leucoanthocyanin glucoside (Steward and Shantz, 1956), chalconaringenin-2′-glucoside (Turetskaya *et al.*, 1968), scopolin (Moreland *et al.*, 1966), and daidzin (Miller, 1969). However, there are some substances which, although inactive when in glycosidic linkage, are rendered very active upon degradation of the glycoside. The glycoside, amygdalin, conspicuous in the seeds of stone fruits, yields toxic hydrogen cyanide (HCN) upon hydrolysis (see Conn and Butler, 1969, and reaction 7.1).

$$C_6H_5-\underset{H}{\overset{O-\beta\text{-gentiobiose}}{C}}-C\equiv N \xrightarrow{\text{enzyme}} C_6H_5-\overset{O}{\overset{\|}{C}}-H + HCN \quad (7.1)$$

Amygdalin Benzaldehyde

Members of the mustard family (Brassicaceae) contain the so-called mustard oil glycosides (Kjaer, 1960). These yield upon hydrolysis volatile oils of mustard (see reaction 7.2). Allyl and β-phenethylisothiocyanate are only two examples of compounds of this class that act as germination and growth inhibitors (Ettlinger and Kjaer, 1968).

$$R{-}C({=}NOSO_3^-)({-}S\text{-glucose}) \xrightarrow[H_2O]{\text{myrosinase}} [R{-}N{=}C{=}S] + \text{glucose} + \text{sulfate} \quad (7.2)$$

$$[R{-}N{=}C{=}S] \rightarrow ROH^- + SCN^-$$

Virtanen's group in Finland has demonstrated a very subtle type of resistance to infection by *Fusarium* (Virtanen and Hietala, 1960). Resistant rye plants contain a glucoside that, as such, has no effect on the microorganism; however when the plant is injured, enzymatic cleavage produces the aglucone, 2,4-dihydroxy-1,4-benzoxazin-3-one (see reaction 7.3), which does inhibit the growth of the microorganism. The methoxy derivative [6-methoxy-2(3)-benzoxazolinone] found in wheat and maize is likewise inhibitory (see Sam and Valentine, 1969, for a review of 2-benzoxazolinones; Spencer, 1963).

$$\text{Glucoside } (C_{14}H_{17}O_9N)\ [\text{CHO}\cdot C_6H_{11}O_5,\ \text{CO},\ \text{N–OH}] \xrightarrow{\text{enzyme}} \text{Aglucone (2,4-dihydroxy-1,4-benzoxazin-3-one)}\ [\text{CHOH},\ \text{CO},\ \text{N–OH}] \longrightarrow \text{Benzoxazolinone}\ [\text{O–CO},\ \text{N–H}] + HCOOH \quad (7.3)$$

Glucoside
($C_{14}H_{17}O_9N$)

Aglucone
(2,4-dihydroxy-1,4-benzoxazin-3-one)

Benzoxazolinone

Other glycosides may promote cell division. Immature *Aesculus* fluids have yielded an IAA–glucose–rhamnose complex that is very active, especially in the presence of *myo*-inositol, in the carrot growth assay system (Shantz and Steward, 1968). An active arabinose–IAA complex was previously isolated from *Zea* (see Steward, 1968).

Ethylene and Ethylene-Releasing Agents

Ethylene, a natural regulator of growth in higher plants, is ubiquitous

(Burg and Burg, 1968; Mapson, 1969; Pratt and Goeschl, 1969). 2-Chloroethanephosphonic acid represents a new class of synthetic plant growth regulators which produce a variety of growth responses in plants (Hallaway and Osborne, 1969; Jackson and Osborne, 1970). When it is applied, ethylene is directly released to the plant tissues and can thereby be used to regulate various phases of plant metabolism, growth, and development (see Fig. 7-6) (Cooke and Randall, 1968). But a very large number of plant

$$ClCH_2CH_2\overset{\overset{O}{\|}}{P}\begin{matrix} \diagup OH \\ \diagdown OH \end{matrix}$$

2-Chloroethanephosphonic acid

growth regulators are now known to cause ethylene to be released so that their effectiveness is now being attributed to that substance! The ability of naphthaleneacetic acid to induce flowering in pineapple is an example of a synthetic auxin that has an ethylene-releasing potential. The newer ethylene-yielding compounds are however, in fact, now in routine use to stimulate flowering in major crops, such as pineapple.

Fig. 7-6. Response of scarlet sage (*Salvia*) to a single exposure of 1 day to 20 ppm ethylene. Left, untreated plants; right, treated plants. Note the epinasty and abscission of leaves, suppression of apical growth followed by the abnormal growth of axillary buds. (Photograph courtesy of the U. S. Department of Agriculture.)

Phthalides

Seeds of some plant species (especially of the carrot family, Umbelliferae) contain germination inhibitors that are phthalides (Moreland *et al.*, 1966).

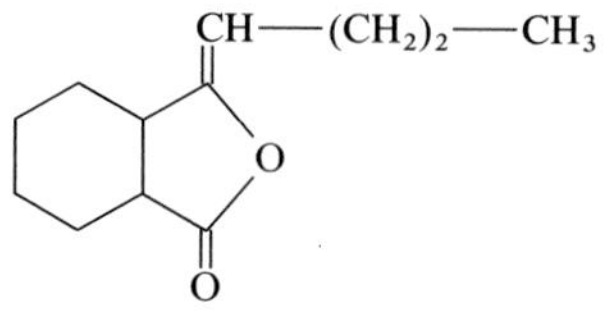

$CH{-}(CH_2)_2{-}CH_3$

O

O

3-*n*-Butylidenehexahydrophthalide

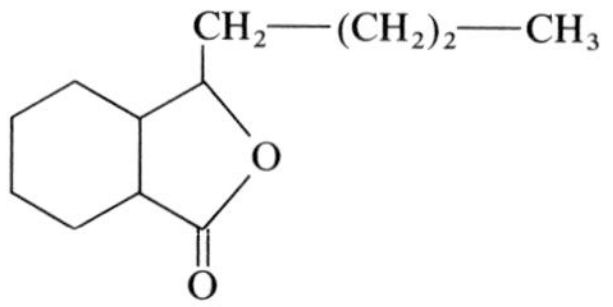

$CH_2{-}(CH_2)_2{-}CH_3$

O

O

3-*n*-Butylhexahydrophthalide

Alkyl Esters

The lower alkyl esters of the C_8 to C_{12} fatty acids and the C_8 to C_{10} fatty alcohols are synthetics that have found use as selective "pruning agents," since they kill or inhibit terminal meristems, whereas axillary meristems remain uninjured (Cathey *et al.*, 1966; Cathey and Steffens, 1968; Steffens and Cathey, 1969).

Fatty Alcohols

Fatty alcohols with chain lengths of C_9, C_{10}, and C_{11} (1-nonanol; 1-decanol; 1-undecanol) are very active in inhibiting axillary and terminal bud growth (Steffens *et al.*, 1967) (see Fig. 7-7).

Aldehydes

Substances such as citral, cinnamaldehyde, salicylaldehyde, and benzaldehyde may be inhibitors of growth and germination (Evenari, 1949, 1961; Grümmer, 1961; Moreland *et al.*, 1966; Toole *et al.*, 1956).

Fig. 7-7. "Chemical pruning" of *Forsythia.* Left, growth on a typical shoot after manual pruning; center, control shoot, unpruned, after growth of the apex; right, treated shoot, sprayed with an emulsion of a fatty acid ester to selectively kill the apex. (Photograph courtesy of the U. S. Department of Agriculture.)

Other Classes of Compounds

It must be apparent by now that there are very many classes of chemicals that can act as plant growth regulators. There are many naturally occurring compounds which have given clues to stimulate the organic chemist to synthesize related compounds which show similar activity (Schuetz and Titus, 1967). Others are completely unrelated and are the results of empirical screening tests. Countless compounds are to be found in the industrial registers.

While the lists presented here make no pretense at being complete, the major categories of compounds have been so mentioned as to facilitate further reading.

As one would expect, new growth-regulating compounds are constantly being disclosed; phaseolic and lunularic acids are examples. Phaseolic acid is a hydroxylated ketocarboxylic acid that has been isolated from

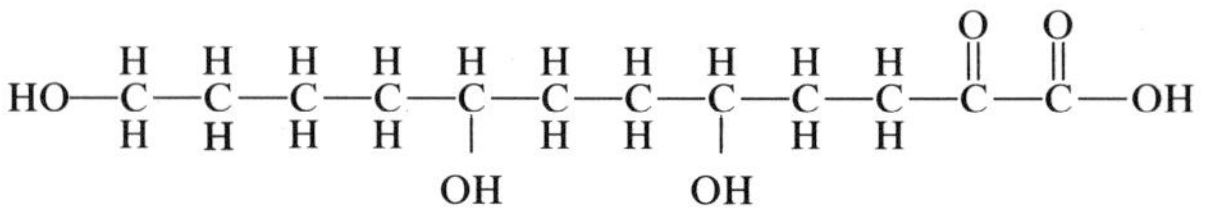

```
                                                      O   O
     H   H   H   H   H   H   H   H   H   H   ||  ||
HO—C—C—C—C—C—C—C—C—C—C—C—C—OH
     H   H   H   H   |   H   H   |   H   H
                    OH          OH
```

or

```
                                                      O—O
     H   H   H   H   H   H   H   H   H   H   ||  ||
HO—C—C—C—C—C—C—C—C—C—C—C—C—OH
     H   H   H   H   |   H   H   |   H   H
                    OH          OH
```

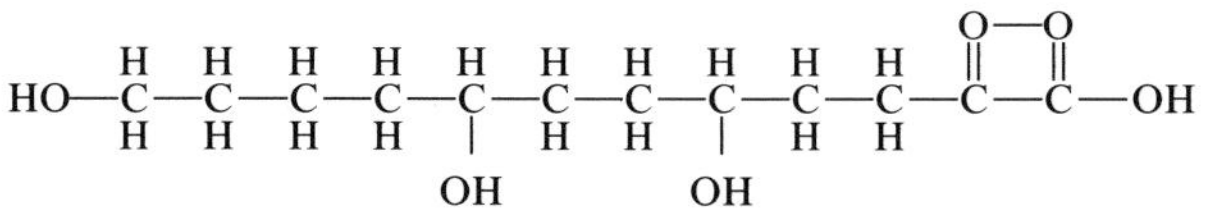

Phaseolic acid

beans (*Phaseolus vulgaris* cv. "Kentucky Wonder"). It shows gibberellin-like activity in that it stimulates the elongation of dwarf maize mutants, although not to the same extent as gibberellic acid; it also induces α-amylase synthesis in the barley endosperm assay, etc. (Redemann *et al.,* 1968). Although phaseolic acid is naturally occurring, the amounts required for its activity are nevertheless high. Lunularic acid (dihydrohydrangeic acid) has been isolated from thalli of the liverwort, *Lunularia cruciata.* This substance is inhibitory to the growth of *Lunularia* gemmae and is active in a number of bioassays (Valio and Schwabe, 1970). Moreover, its concentration varies *in situ* according to day length. Abscisic acid was not detected in the extracts that yielded lunularic acid.

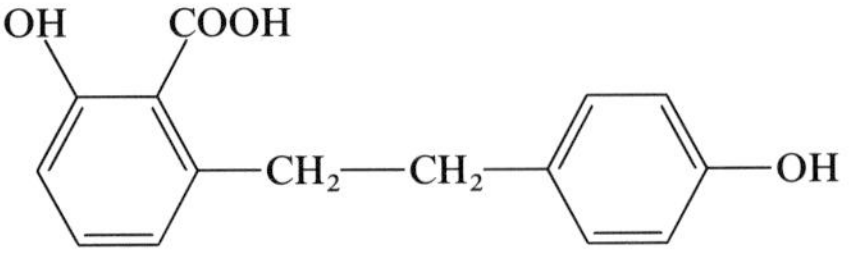

```
OH     COOH
(benzene ring)—CH2—CH2—(benzene ring)—OH
```

Lunularic acid

Methyl 4-chloroindoleacetate, an isolated auxin from immature pea seeds, is of special interest since it is the first example of a naturally occurring chlorine containing plant growth regulator (Marumo *et al.*, 1968).

Substances such as sorbose, deoxycholate (Tatum *et al.*, 1949), and phenylboric acid (Mathan, 1965) may also be added to the list. These also induce morphological effects and should therefore be regarded as plant growth regulators.

The scope and range of action of the substances referred to here obviously defy facile structure/activity interpretations. But, as indicated earlier, the varied events in growing cells and the various ways in which the attributes of organs are correlated in growing plants provide a great array of sites at which growth regulators may act. Within a given category of similar substances that may all be expected to act at the same site and in the same way, feasible general rules that relate physiological activity and chemical structure have been formulated (Crozier *et al.*, 1970; Fregda and Åberg, 1965; Leonard *et al.*, 1968, 1969; Porter and Thimann, 1965; Skoog *et al.*, 1967; Wain and Fawcett, 1969). But when one moves from one similar category of growth regulators to another, new rules need to be formulated and, consequently, there is no all-embracing comprehensive theory.

Growth-regulating substances are still recognized, often empirically, by what they do in given test systems. While a given response is generally attributed to an applied molecule per se, it may in fact rarely act alone. The very fact of its specificity requires some equally defined molecular conformation at the active site. But, since the responses elicited are so general, in contrast to the animal hormone effects which are so specific, one may recognize that many substances now regarded as of slight interest as plant growth regulators could well emerge as such if they were tested under such circumstances that ancillary responses, otherwise limiting, could be simultaneously controlled. This opens up vistas of chemical control and regulation not yet explored, for to do so would virtually require a complete knowledge of all that goes on in any given growing system and of all the ways in which its performance could be regulated by rate-limiting steps.

CHAPTER 8 What Do the Growth-Regulating Substances Do?

GENERAL APPROACHES

Explanations of growth regulation are being sought in mechanisms which will invoke the "turning on or off" of genes and the transmission of the consequential effects via RNA's and the synthesis they activate (see Brachet, 1968; Davidson, 1969; Key, 1969; Key and Ingle, 1968; Loening, 1968; van Overbeek, 1966; van Overbeek *et al.*, 1967; Stern, 1966; Stern and Hotta, 1968). All this requires that every growth-regulating action must occur very specifically at a nuclear site and invoke influences upon nucleic acids and presumptively act at some particular stage of the cell cycle. These lines of approach result from work done on simple systems, e.g., bacteria and viruses. Despite current trends, the validity of this approach should not be accepted without question (see Commoner, 1968).

The penalty for an alternative generalized approach, like the one presented, is the seeming lack of definition or specificity in assigning precise roles to the growth-regulating substances in question. *But the truth is that this is precisely where the subject now stands.*

Indole-3-acetic acid (IAA) still lacks a precisely specified single function or a precisely specifiable site for its action. In some situations the IAA molecule fulfills the role of an auxin, whereas in others, as part of a more complex system, it could be held to function as a cytokinin. In the early period, the action of IAA was thought to be secondary to the more potent effect of certain complex lactones (auxins A and B), which long figured in the plant physiological literature, despite the fact that they were unconfirmed (see Boysen Jensen, 1936). The actual substances which were the basis of the identification of auxins A and B have proved to be, on reexamination, mixtures of well-known, simple substances (Vliegenthart and Vliegenthart, 1966), whereas the supposedly complex molecules, when synthesized, (Hwang and Matsui, 1968) have proved to be inactive (Nakamura *et al.,* 1966).

IAA has variously had ascribed to it the ability to evoke numerous responses, from "plasticising" cell walls (Cleland, 1965, and 1968; Cleland *et al.*, 1968; Datko and Maclachlan, 1968; Masuda, 1968; Ray, 1969; Wilson, 1964) so that preformed cells may expand, to its claimed ability to interpose its effect when a flash of light on leaves counteracts the exposure to a long night, with consequential and profound effects upon the behavior of the growing shoot (Bonner and Liverman, 1953).

Over the years, IAA has been invoked in one way or another at almost every conceivable biochemical and cell physiological level. These involvements have included the active internal secretion of water and solutes, cytoplasmic streaming, cellulose and pectin changes in developing walls, effects on respiration mediated through enzymes as of the Krebs cycle and the terminal oxidase system, the induction and repression of peroxidases, the promotion of DNA synthesis, the stimulation of mRNA's with consequential increases of specific proteins and enzymes, and recently the involvement of IAA in synthesis and release of ethylene as a physiologically active agent (see Galston, 1967b; Galston and Davies, 1969, 1970; Galston and Purves, 1960; van Overbeek, 1966; Thimann, 1969). It is obviously impossible here to deal with each of these suggestions in their historical context, especially since none of these approaches have achieved finality.

Although their role in protein and enzyme synthesis has been invoked for many growth regulators (Koblitz, 1969a,b), it has been more specifically investigated in a system (the barley aleurone bioassay; see Fig. 4-5) involving the gibberellins. In this system, given gibberellic acids induce the *de novo* synthesis of protein molecules in the form of enzymes, e.g., amylases, proteases, etc. (Filner *et al.*, 1969; Filner and Varner, 1967; Varner and Johri, 1968). Whereas this may explain the role of gibberellins in germinating grains, this is not necessarily the way they would also act in other systems, e.g., the elongation of otherwise dwarfed shoots (Brian and Hemming, 1955; Phinney, 1956). [In fact, fumaric acid, as secreted by an insect, has been shown to have the same effect as a gibberellin on rice endosperms (Tamura *et al.*, 1967).]

Perhaps the most dramatic example of a family of growth responses to a hormone is that due to ecdysone on insect larvae and the correlation of its effects with visible "puffs" of the giant salivary gland chromosomes (see Beermann, 1966; Gall, 1963; John and Lewis, 1968, and references cited therein). Dramatic as this evidence seems to be, it always raises the question whether the localized RNA activity in the form of the "puffs" demonstrates the site of ecdysone action or merely describes its morphogenetic consequences at another level. Moreover, a wide range of other compounds such as simple salts induce the equivalent chromosomal effects (see Kroeger, 1967).

If growth is modulated by activated genes, and if the genes are regulated by specific nuclear histones (see Bonner, 1965; Bonner *et al.*, 1968, for the early views of this) any growth-regulating substance should mediate its effects via histones, but this merely transfers the problem to the means of control of the histones. Organ specific histones have not been found, and the suggestion that the "same few histone fractions may repress different genes in different cells" (Fambrough *et al.*, 1968) has been complicated by the finding that histones with different base composition can be equally effective in blocking template activity of the same DNA (Shih and Bonner, 1970). Some have taken the view (see Pogo *et al.,* 1966) that "a change in the structure of the chromatin—brought about by or coincident with acetylation of histones—is a necessary prerequisite to the synthesis of new RNA's at previously repressed gene loci" (see Clever and Ellgaard, 1970, and references cited therein). However, others have shown that "a stimulation of RNA synthesis, at previously repressed loci, does not generally involve a prior acetylation of histones" (see Clever and Ellgaard, 1970). These workers even go so far as to raise the view that "if it should be shown that processes involving histone acetylation are indeed a necessary prerequisite for the initiation of transcription," then they would "have to consider the unorthodox but interesting possibility that puff formation may not result from a local stimulation of RNA synthesis" (Clever and Ellgaard, 1970). This being so, from where does the specificity in the supposed histone regulation of gene activity derive?

Some have suggested that synthetic cytokinins like N^6-substituted adenines, with diverse structures, may mimic the natural corepressor in a Jacob–Monod system (see Fig. 8-1) and so release the operator site, which in turn activates a structural gene to mediate protein synthesis (Jacob and Monod, 1963). Reasonable as all this may seem to be for the adenyl derivatives, it could hardly embrace the effects of the great diversity of molecules now known to stimulate cell division in plants. Braun and Wood attribute (Wood *et al.,* 1969) the cytokinin effect equally specifically but to a different type of molecule, e.g., to certain purinone derivatives, which are held to act, not at a synthetic part in the cell cycle for "they are directly involved in promoting cytokinesis" (M) (see Fig. 2-8). Braun believes that the other substances to which cytokinin activity has been attributed act because they can promote the synthesis of the purinone derivatives and thus intervene at a different stage of the cell cycle. We shall return to this point again.

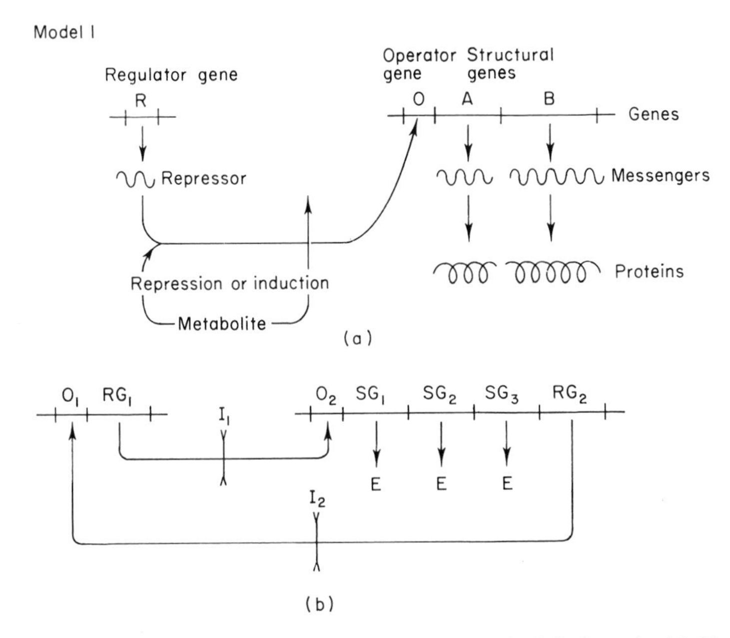

Fig. 8-1. General models for the regulation of enzyme synthesis in bacteria. (a) *The simple operon model.* R, The regulator gene; the repressor associates with some inducing or repressing substance shown here as a metabolite; O, the operator gene; structural genes are represented as A and B, etc.; messengers are made by structural genes; proteins are made by ribosomes associated with their messengers. The above concept now includes the possibility that the repressor may be a protein, itself subject to gene control (Gilbert and Müller-Hill, 1966). On this view, the cytokinins also could intervene at the repressor or induction site and so control the operator O. (b) *Compound operon system.* O is the operator; E is the enzyme synthesized. The regulatory gene RG_1 controls the activity of an operon containing three structural genes (SG_1, SG_2, and SG_3) and another regulatory gene RG_2. The regulatory gene RG_1 itself belongs to another operon sensitive to the repressor synthesized by RG_2. The action of RG_1 can be antagonized by an inducer I_1, which activates SG_1, SG_2, SG_3, and RG_2 (and therefore inactivates RG_1). The action of RG_2 can be antagonized by an inducer I_2 which activates RG_1 (and therefore inactivates the systems SG_1, SG_2, SG_3, and RG_2). [From Jacob and Monod (1963); see also Beckwith and Zipser (1970).]

It often seems easier to be definite about the role of substances that inhibit growth or have herbicidal action (Moreland, 1967), for it may only require one limiting metabolic event for all growth to stop, as in the role of CMU [3-(*p*-chlorophenyl)-1,1-dimethylurea] which inhibits the Hill reaction of photosynthesis (Izawa and Good, 1965). Similarly, phloretin, the aglucone of phloridzin inhibits photophosphorylation by interacting with both the transphosphorylating and electron transport systems (Uribe, 1970). Colchicine inhibits the proteins that comprise the spindle of the mitotic apparatus by binding to the protein subunits of the microtubules.

In this way, the completion of nuclear division is prevented (see Deysson, 1968; Ilan and Quastel, 1966, and references cited therein; Shelanski and Taylor, 1967).

Each trend of thought suggested by the modern molecular biology doctrine, which attempts to attribute the sole growth regulatory role to a given molecule, or class of molecules, ends in the same dilemma. Namely, there are too many different ways by which the growth may be modified and there is too little evidence that any specific mechanism actually works.

Controls of Growth versus Controls of Metabolism

The difficulty of assigning roles and sites of action to growth regulatory substances is indicated by the equal difficulty of simulating, in culture, the biochemical events which occur in cells *in situ* that obviously have "built-in" capacities that are not expressed in culture. Indeed, if growth regulators prove to be the ready means to "tell" cultured tissue to "start acting like cells and tissue *in situ*," then a major new industry will emerge, for it ought then to be feasible to culture the tissues of organs with specialized biochemistry and to induce them, at will, to form the typical compounds in question. These problems have been examined from the standpoint of the formation in cultured cells of special proteins (enzymes), of special metabolites, e.g., of nitrogen compounds, of commercially valuable substances such as essential oils or flavors, drugs, e.g., alkaloids, or of other substances which accumulate in various cells or tissues (Krikorian and Steward, 1969). However, even such an apparently simple problem as to induce cultured, free potato cells to form the large starch grains of the tuber cells, or cultured, free carrot cells to form the bright orange-red carotenes of the intact and senescent tissue, have not as yet been solved (Israel *et al.*, 1969). Although the appropriate degree of control over such cells should eventually emerge, the bulk of present evidence indicates a current inability to instruct the unorganized cells to achieve a given result, outside the milieu in which they are formed and within which they function spontaneously. All this may be dismissed as merely a problem of activating, or derepressing, the appropriate set of genes, and of reversing that action when so desired. Those who see every regulatory role in terms of specific protein or enzyme syntheses must recognize a great range of growth factors, which must all be able to intervene in cells as they combine with or modify ribonucleic acids. But is it chemically feasible that all these diverse molecules [from ethylene (Shimo-

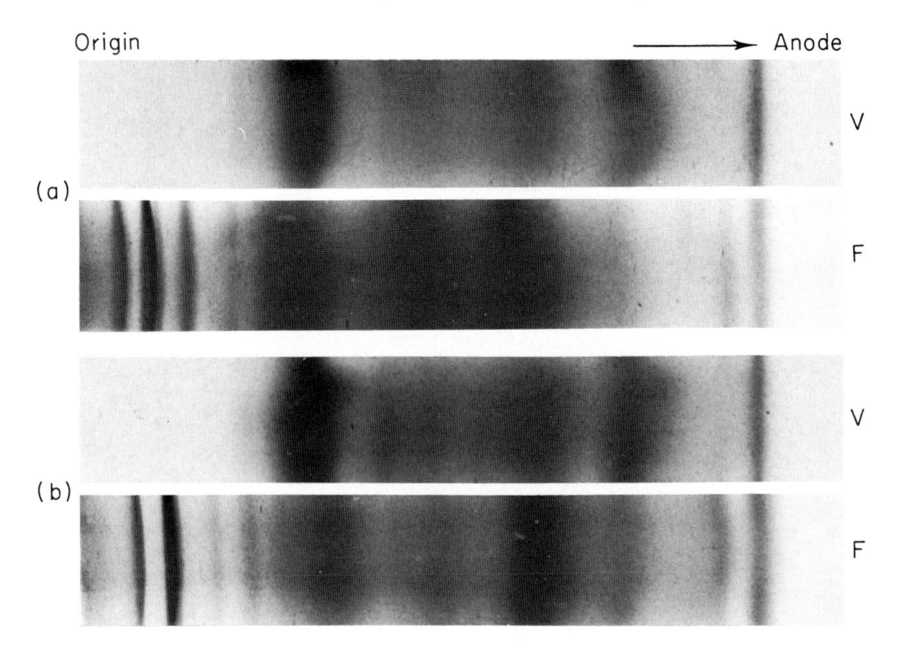

Fig. 8-2. The signals to the tulip apex change its biochemistry. Soluble proteins extracted from growing points were separated on acrylamide gels from two cultivars (a and b) and from vegetative (V) and florally induced (F) bulbs. (The gels were deliberately overloaded to emphasize the contrast for those bands closest to the origin.) [From Barber and Steward (1968).]

kawa and Kasai, 1968) to phytochrome (Furuya, 1968) to the gibberellic acids] can successfully converge upon, and act as they combine with or modify, the nucleic acids of the cells they affect (Key, 1969; Trewavas, 1968a, b)?

The control of biochemical potentialities which accompanies normal plant development has been detected, though not explained, by the techniques of gel electrophoresis. This technique was first applied to protein gradients in pea seedlings, but more comprehensive studies have been made on the tulip (Barber and Steward, 1968). A particular temperature-mediated control over floral induction in the growing points of otherwise vegetative tulip bulbs (see Fig. 1-5) also mediates some recognizable, even dramatic, changes in the protein and enzyme complements of their growing points (see Fig. 8-2). If the effect of the temperature stimulus is perceived elsewhere, and is transmitted chemically to the growing shoot tip, we have still no knowledge of the "messenger" molecule so concerned, despite the fact that they have been termed "vernalins" (Lang, 1965; Melchers and Lang, 1948). But the lesson to be learned from this sort of work is the following.

Under the genetic plan, organs develop. Constituent cells retain their unique genetic constitution, but each organ furnishes a milieu which limits, or controls, its expression. Biochemical synthesis seems to follow the lead of the morphology rather than the reverse. Work on a single gene mutant (peak) in *Neurospora crassa,* which constantly affected the protein patterns, along with the form of the culture, showed that *any* factor, genetic *or* phenotypic, which controls the morphology also dictates the protein complements (Barber *et al.*, 1969). But the distinctive effects of genetically different strains could be overcome by chemical agents (sorbose or deoxycholate), which rendered genotypically distinct strains phenotypically similar (Tatum *et al.*, 1949). A potentially similar case in higher plants is the role of phenylboric acid, which affects the leaf form and simulates the effect of the "lanceolate gene" in tomato (Mathan, 1965).

Embryological Controls: Plants versus Animals

The stores of substance available in fertilized amphibian eggs permit their very rapid development. After fertilization, cell divisions occur very rapidly. According to Gurdon (1968), the eggs of *Xenopus* may pass through cleavage in less than 8 hours, during which some 12 to 15 cell generations produce about 15,000 cells. Thereafter, morphogenesis is equally rapid (a young tadpole is formed in a further 18 hours). Throughout this period the course of nucleic acid synthesis (DNA and classes of RNA) has been traced (Gurdon, 1969). The trends in the levels of nucleic acids so observed constitute the biochemical evidence for the high degree of regulation in the developing amphibian embryo. Even so, the nucleic acids present reflect the activity of genes; they do not tell how they are activated. Through this period, the amphibian egg behaves in a very predetermined fashion; it follows out very time-dependent events which are already decreed within the egg. It is as though the egg cytoplasm were "telling the nucleus what to do" without any intervention of exogenous controls of the sort that may be mediated by growth substances in the morphogenesis of free plant cells. If a plant parallel is to be sought, it should be with the germination of a seed, in which an embryo and its storage material contain the information and the needed substance for the first events which follow in rapid succession. However, Gurdon's final conclusion still reflects his uncertainty "whether a given regulatory mechanism does in fact operate in living cells," for he recognizes that "the activity of the same gene may be regulated in different ways under different conditions."

Stimuli and Response: Levels of Manifestation

Growth-regulating substances may affect any, or all, of the main events that growth entails. These include such processes as: (a) the absorption of water and solutes which develop the cell's ability to build its surface and maintain its turgor and increase its volume (Steward and Mott, 1970); (b) the formation of new cell walls and provision for their elongation; (c) the synthesis of large molecules, and especially the replication of DNA itself, which culminates in the formation of "self-duplicating" structures (Watson, 1970). It is current practice to focus attention upon some single biochemical event, and hopefully seek by this means a causal explanation. But the observed effects of growth regulators on cells involve the synthesis of many of these events, as in cell enlargement on the one hand and cell division on the other (see Heslop–Harrison, 1967), or in the subsequent onset of maturation, dormancy, and senescence of cells (Woolhouse, 1967). Even if all was known about the relations between specific growth factors, their relations to DNA and to the consequential proteins synthesized, there would still be a great gap between this knowledge and the manner in which the morphogenetic effects (emergence of shape and form) of cells and organs are achieved (see Green, 1969).

There seems to be no generally accepted evidence that ascribes the observed effects to their underlying causes at specifiable cell sites. We only see the dramatic and very delayed actions that result. In an ideal situation we would be able to refine the techniques of fine structure and of radioautography so as to locate the target sites and those of high-speed reaction kinetics to distinguish the primary acts from their subsequent consequences (see Villiers, 1968; Zwar and Brown, 1968). It should not be surprising that so few ultimate explanations of modes of action are available, for we are still describing at different levels, the effects which are produced. Here one may merely refer to the published report of the International Conference on Plant Growth Substances held in 1967, which did not in fact appear until mid-1969 (Wightman and Setterfield, 1968). This volume of 1642 pages contains 115 papers; it tells much about plant growth regulators and their miscellaneous effects in different situations. But remarkably little finality is reached about how, or where, these substances act to regulate the development of any given cell (whether it be a zygote or a somatic cell) or growing region (vegetative or reproductive) within limits set by the genetic information available to each of the living, totipotent cells. But, insofar as carrot explants and their cells are concerned, the effect of growth-regulating substances and systems may be seen at different levels and by very different means (Steward *et al.*, 1968a).

The complete exogenous growth factor system which makes carrot cells grow rapidly produces visible effects at each fine structural level of scrutiny. Mitochondria become elongated and coarsely structured internally, and these features are seemingly more characteristic of metabolically active than of quiescent cells (Israel and Steward, 1966). Plastids, which in quiescent, secondary phloem cells are deficient in internal structure although they contain rich red carotene, become mature chloroplasts in the light after a sequence of dramatic morphogenetic changes (Israel *et al.*, 1969; Israel and Steward, 1967). Although ribosomes (and their contained nucleic acids) become very abundant in ground cytoplasm, even in metabolically activated cells on a basal nutrient medium, they seem to lack the signals to make them biochemically effective (Israel and Steward, 1966; Steward *et al.*, 1964a). A curious metabolically stable, hydroxyproline-rich component of the alcohol-insoluble nitrogen (protein) fraction, which seems to become prominent in the renewed activity of otherwise quiescent cells as they prepare to grow, arises *in situ* from exogenous proline (Pollard and Steward, 1959; Steward and Pollard, 1958). But if the formation of such a stable structural moiety, thought to be diffused through the cytoplasm (Israel *et al.*, 1968), is one observed feature of reactivated cells, there is also a renewed general protein synthesis and activated turnover to be considered as another.

Cellular Compartmentation and Integration

The setting in which the growth regulators act is to be seen in terms of the highly compartmented protoplasm, with its numerous vesiculate inclusions. Even amino acids combined in tRNA complexes may be regarded as sequestered, or compartmented. Plant cells are able to isolate, in this way, a great deal of their soluble metabolites from the active centers of synthesis (Steward and Bidwell, 1966). The amino acid molecules which are en route to protein may not mingle with their storage counterparts elsewhere in the cell. Among the very large number of nonprotein nitrogen compounds now known in plants, there are many that may function as metabolic inhibitors and suppress reactions, e.g., of protein synthesis, that are needed in growth. For example, azetidine-2-carboxylic acid, a proline analog, is a competitive proline inhibitor which, like many of its analogs, suppresses growth of cultured carrot explants (Steward *et al.*, 1958c). The reason is that, the azetidine molecule is smaller than its proline counterpart and can thus enter (perhaps via a tRNA) into the configuration of a then unusable protein. A larger proline analog, namely, pipecolic acid, is too big to do this and does not act competitively with proline.

The general picture of cellular compartmentation, enforced by the study of nitrogen metabolism, has a corollary in the evidence that the systems which stimulate growth in the cells (whether these are or are not termed cytokinins) foster *both* synthesis and turnover, and they also permit communications between otherwise isolated pools of metabolites to be bridged by selected molecules. In carrot cells, for example, alanine seems to link the amide-rich storage pool of nitrogen compounds to the sites of protein formation where nitrogen must be donated to carbon frameworks, derived from sugar, in the synthesis of the amino acids which are en route to protein (see Steward, 1968, and references cited therein).

If proteins and enzymes motivate the "unit-processes" of metabolism, if the phosphoric anhydrides represent the universal currency through which the "energy costs" are transacted, if the organelles act as compartments (often autonomous) comparable to "workshops" in which sequences of specialized reactions are performed and specialized products are fabricated, then the growth-regulating substances and systems emerge as the controlling agents or messages ("activators" or "paralyzators" in the Blackman phrase) (see page 72), which preserve, or change, the balanced operation of the whole. In this context the surprise is not that there are now so many exogenous substances and combinations that are involved, as that it could ever have been considered credible that there would only be few! Since there are so many exogenous morphogenetic stimuli, there must also be many autonomous sites for their action. The nucleus clearly transmits at each division the information that determines the range of feasible events which the cell and its discrete organelles can perform, but it seems neither necessary nor credible that it is directly involved in the determination of every subsequent event.

SPECIFIC APPROACHES

More specific interpretations have been offered to explain the role of substances that function as cytokinins in stimulating cell divisions. The cytokinins in question are thought to act because they mediate protein synthesis (Galston and Davies, 1969, 1970; Helgeson, 1968; Skoog and Armstrong, 1970; Srivastava, 1967b; Xhaufflaire and Gaspar, 1968), (see Fig. 8-3). In general, this need not be questioned, but the main point is that the proteins concerned must be special ones. To work, this mechanism must surely involve the synthesis of some highly specific protein vital to

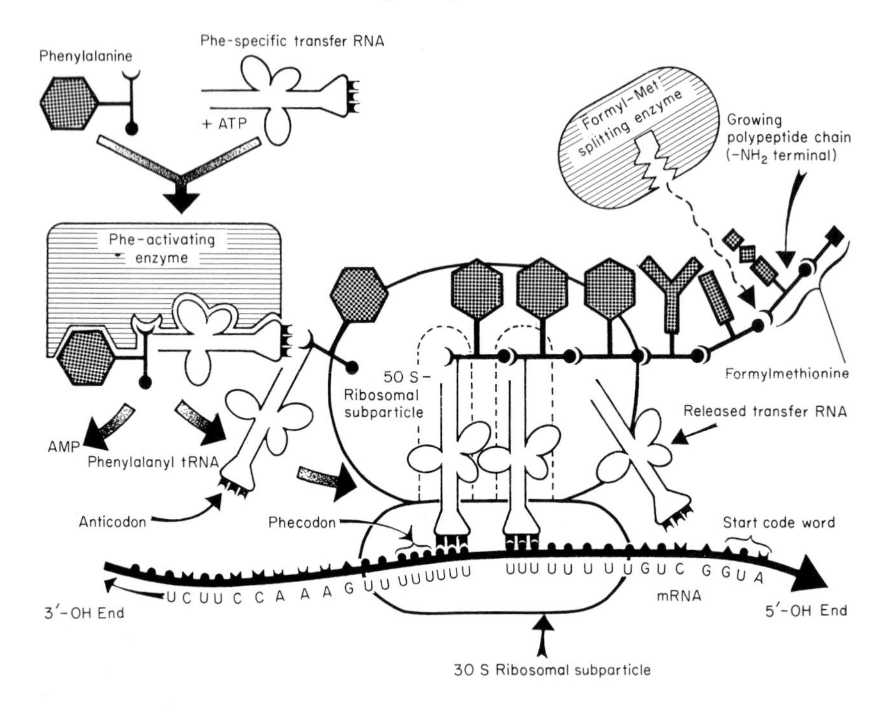

Fig. 8-3. Schematic mechanism of protein synthesis. The synthesis of a hypothetical polypeptide with the amino acid sequence formylmethionine–alanine–valine–phenylalanine–phenylalanine–phenylalanine–phenylalanine ... by *E. coli* is illustrated. The amino acid about to be incorporated is phenylalanine (see upper left) which reacts specifically with phenylalanine–tRNA forming the amino acid ester (phenylalanyl–tRNA). This reaction is brought about by an activating enzyme, as shown. The released phenylalanyl–tRNA is then bound to the larger ribosomal subparticle (50 S), but it is able to recognize the triplet of nucleotides (the codon UUU) of the mRNA, through the anticodon AAA (i.e., the phenylalanine "code word" on the figure). The NH_2 group of the phenylalanyl residue (shown schematically) of the phenylalanyl–tRNA forms a peptide bond with the COOH group of the last amino acid of the growing polypeptide chain, which becomes attached to the phenylalanyl–tRNA. The genetic message should be read from the 5′ end to the 3′ end, i.e., from right to left. The codon AUG marks the start of the message, and the codons UAA and UAG may signal the release of the synthesized polypeptide chain. When the ribosome reaches the 3′ end of the mRNA, the ribosomal particle may fall apart (into 50 S and 30 S subparticles). The whole process may then be repeated as ribosomes are reconstituted and attach to the mRNA. The larger ribosomal subparticle (50 S) may then react with the smaller ones (30 S) to rebuild the ribosome attached to the mRNA, and presenting, again, the site for peptide bond synthesis (3′ end of the tRNA) to the activated amino acid. [By permission of The Chemical Society, London; from Cox (1968).]

cell multiplication, perhaps an enzyme like DNA polymerase, but for which the evidence may not yet be available. These ideas integrate with modern molecular biological thought as follows.

The Role of Adenyl Compounds

The isolation and activity of kinetin from old or autoclaved DNA from herring sperm (Miller *et al.*, 1955a) early suggested that it might justifiably be implicated in some way with nucleic acid metabolism. This possibility came to the fore when it was shown that kinetin induced DNA synthesis in tobacco cultures in the presence of auxin, as indeed one would expect since the cells divided. As early as 1960, Thimann and Laloraya suggested that kinetin could possibly be incorporated into nucleic acid and that this nucleic acid could, in turn, be the basis of increased protein synthesis. Other workers conjectured that unusual or abnormal nucleic acids might be synthesized in response to kinetin, but it was soon shown that there were no appreciable differences either in profile, nucleotide composition, or a variety of chemical and physical properties between kinetin-treated and control tissues (Srivastava, 1967a). As the techniques for the study of nucleic acids became more refined, however, it seemed that the growth regulatory substances influenced the kind, as well as the amount of RNA (see Key, 1969).

Early experiments to test the incorporation of kinetin-8-^{14}C into tobacco callus cultures and into barley and tobacco leaves showed that although much of it was metabolized, radioactive carbon was recovered in all the nucleic acid fractions examined (e.g., 19.5% in tRNA, 6.5% in DNA, and 73.9% in ribosomal RNA in the case of the barley leaves, and 19.1% in tRNA, 11.6% in DNA, and 69.1% in ribosomal RNA in the case of cultured tobacco callus) and hence were essentially inconclusive (Srivastava, 1967a). The work of Fox, however, which utilized 6-benzylaminopurine-8-^{14}C and 6-benzylaminopurine-methylene-^{14}C, suggested that the cytokinins could be incorporated preferentially into certain soluble RNA's of cultured tobacco and soybean, even though the bulk of the material supplied was metabolized (Fox, 1966, 1969). The main deficiency of this work, as the investigators themselves recognized, was that the radioactive material

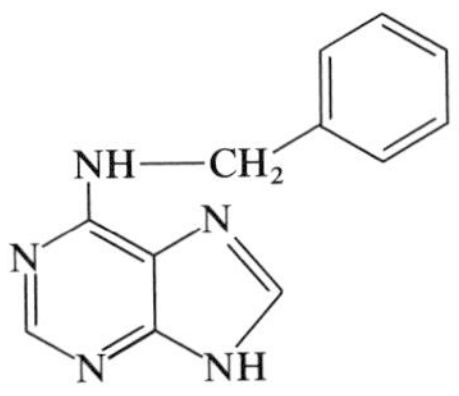

N^6-Benzyladenine

recovered may not have been a nucleotide, may not have been bound covalently into RNA, or may not even have been benzyladenine. It was clear from the work, however, that the radioactivity in whatever compounds, was not associated with ribosomal RNA (Fox, 1966).

Adenyl Cytokinins in tRNA

From these results, Fox hypothesized that by virtue of their possible incorporation into tRNA's, cytokinins provided the biological equivalent of RNA methylation (a reaction thought to confer amino acid transfer competency) in those plants in which the genetic information for the synthesis of RNA-methylating enzymes had been "switched off" during differentiation (Fox and Chen, 1967). Of course, it is obvious that this view failed at that time to explain just how the cytokinin(s) could replace, functionally, the various methylated bases that RNA's contain even if it were so incorporated.

Progress in the rigid chemical determination of the primary sequence of the bases of various tRNA species gave special impetus to this work (see Zachau, 1969). It permitted the idea that indeed cytokinins were part of at least some tRNA's. Holley *et al.*, (1965) were the first to report the presence of an unusual base immediately adjacent to the anticodon, the anticodon being the base sequence which carries the complementary code for the messenger RNA which, in turn, specifies the appropriate place for the amino acids in the new protein being synthesized. The new base was rigidly identified as N^6-(Δ^2-isopentenyl)adenosine (see formula for ring numbers) [γ,γ-(dimethylallyl)adenosine] (Zachau *et al.*, 1966) (see Fig. 8-4).

There are a number of chemically established tRNA primary structure sequences (see Dunn and Hall, 1970; Zachau, 1969). A number of these have been critically shown to contain modified nucleoside bases which

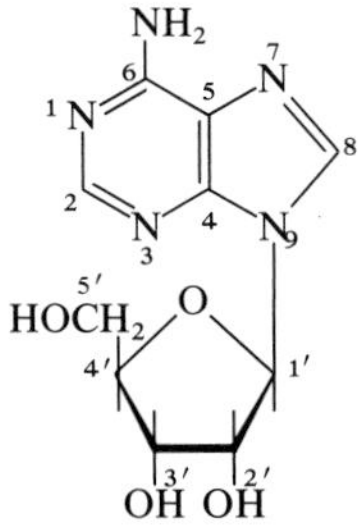

Adenosine

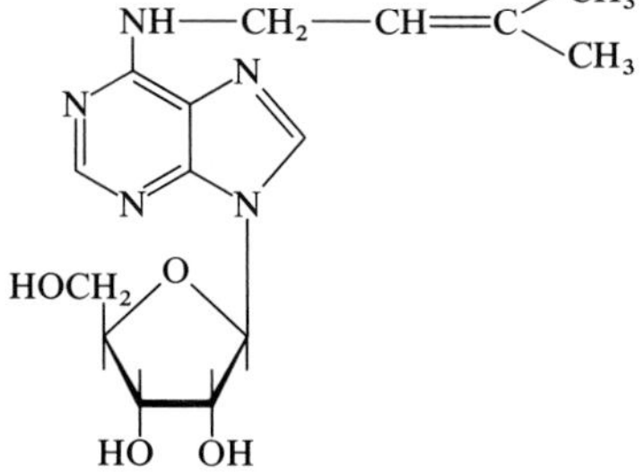

N^6-(Δ^2-Isopentenyl)adenosine

when free have cytokinin activity. Although these bases have not yet been precisely located in a purified tRNA from a higher plant (see Table 8-1), there is still an incompletely identified fluorescent moiety adjacent to the anticodon of wheat germ phenylalanine tRNA (Dudock *et al.*, 1969), which may well be still another unusual base in tRNA's (see Yoshikami, 1970).

Nucleosides have been found that are designated as "hypermodified," i.e., nucleosides that have (a) a relatively large side chain of four carbons; (b) a functional group such as a hydroxyl, carboxyl, or the allylic double bond in the side chain; and (c) are located adjacent to the 3′ end of the anticodon (see Hall, 1970, and Fig. 8-4). This finding has great importance because of the consequences for the properties and reactivity of the tRNA molecule.

A wide range of organisms yield RNA preparations which possess cytokinin activity in the tobacco pith assay (Klämbt and Kovoor, 1969; Skoog and Armstrong, 1970). From these preparations ribonucleosides have been

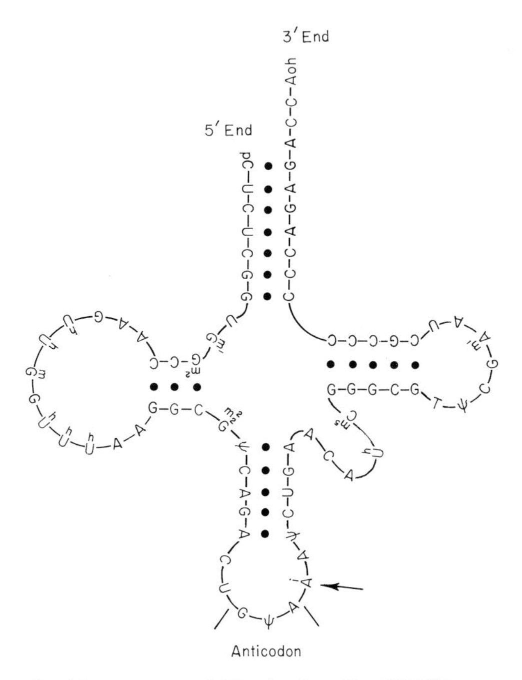

Fig. 8-4. The nucleotide sequence of *Torulopsis utilis* $tRNA^{Tyr}$ arranged in a cloverleaf pattern. The abbreviations used are: p and –, mono- and diesterified phosphate group; A, adenosine; G, guanosine; C, cytidine; U, uridine; ψ, pseudouridine; hU, 5,6-dihydrouridine; T, ribothymidine; I, inosine; m^1A, 1-methyladenosine; iA, N^6-isopentenyladenosine; m^1G, 1-methylguanosine; m^2G, N^2-methylguanosine; m^2_2G, N^2-dimethylguanosine; m^5C, 5-methylcytidine; Gm, 2′-*O*-methylguanosine. Note the position of the N^6-isopentenyladenosine. [After Hashimoto *et al.*, (1969).]

TABLE 8-1
BASE SEQUENCE OF THE ANTICODON AND THE NUCLEOSIDE ADJACENT TO THE 3′ END OF THE ANTICODON IN SOME IDENTIFIED tRNA'S

Source and description[a]	Anticodon (5′ → 3′)	Adjacent group
Saccharomyces cerevisiae		
$tRNA^{Ala}_1$	IGC	1-Methylinosine
$tRNA^{Phe}$	GmAA	Unidentified "Y"
$tRNA^{Ser}_{1\,(2)}$	IGA	N^6-(Δ^2-Isopentenyl)adenosine
$tRNA^{Tyr}$	GψA	N^6-(Δ^2-Isopentenyl)adenosine
$tRNA^{Val}_1$	IAC	Adenosine
Torulopsis utilis		
$tRNA^{Ile}$	IAU	N-(Purin-6-ylcarbamoyl)-threonine riboside
$tRNA^{Val}_1$	IAC	Adenosine
$tRNA^{Tyr}$	GψA	N^6-(Δ^2-Isopentenyl)adenosine
Escherichia coli		
$tRNA^{Met}_f$	CAU	Adenosine
$tRNA^{Met}_m$	C*AU	N-(Purin-6-ylcarbamoyl)-threonine riboside
$tRNA^{Tyr}$	G*UA	N^6-(Δ^2-Isopentenyl)-2-methylthioadenosine
$tRNA^{Val}_1$	–AC	N^6-Methyladenosine
$tRNA^{Phe}$	GAA	N^6-(Δ^2-Isopentenyl)-2-methylthioadenosine
$tRNA^{Leu}$	CAG	1 or 2-Methylguanylic acid
$tRNA^{Trp}$	CCA	2-Methylthio-6-isopentenyl adenosine
Triticum duris (wheat germ)		
$tRNA^{Phe}$	GmAA	Unidentified base like "Y"
Rattus rattus (liver)		
$tRNA^{Ser}$	IGA	N^6-(Δ^2-Isopentenyl)adenosine

[a] Conventions in the table convey additional information for they describe the specific tRNA in question; viz., $tRNA^{Met}_f$ denotes formylmethionine; $tRNA^{Val}_1$, etc., is the isoacceptor designation. [Further details on the nomenclature and symbols, see Dunn and Hall (1970) and the IUPAC-IUB Combined Commission on Nomenclature recommendations (1970).]

A denotes adenosine; G, guanosine; I, inosine; U, uridine; C, cytidine; Gm, 2′-*O*-methylguanosine; m^1G, 1-methylguanosine; ψ, pseudouridine; – represents uridine-5-oxyacetic acid as the 5′ end of the anticodon in $tRNA^{Val}_1$. An asterisk (*) next to a nucleoside indicates it is an unidentified derivative of that nucleoside. $tRNA^{Ala}$, etc., denotes alanine accepting transfer RNA, etc. (Burrows *et al.*, 1969, 1970; Dunn and Hall, 1970; Hall, 1970; Hecht *et al.*, 1969a,b; Zachau, 1969).

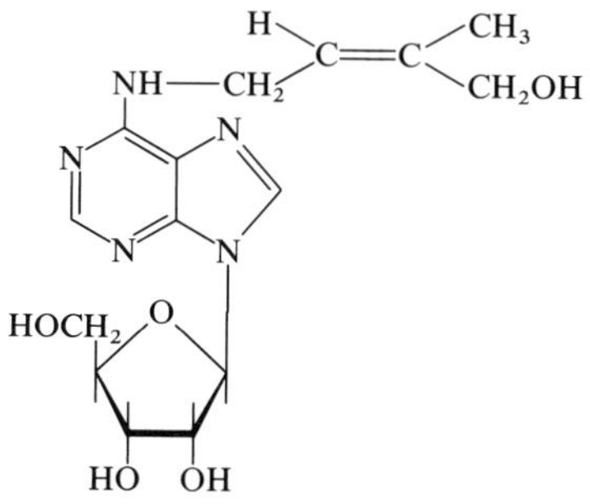

N^6-(Δ^2-*cis*-4-Hydroxy-3-methylbut-2-enyl)adenosine

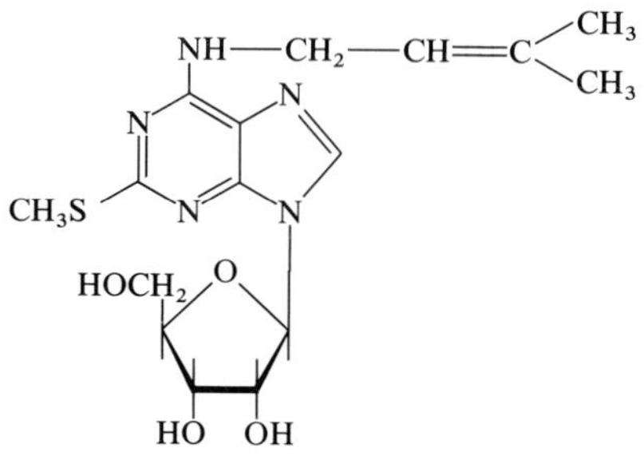

N^6-(Δ^2-Isopentenyl)-2-methylthioadenosine

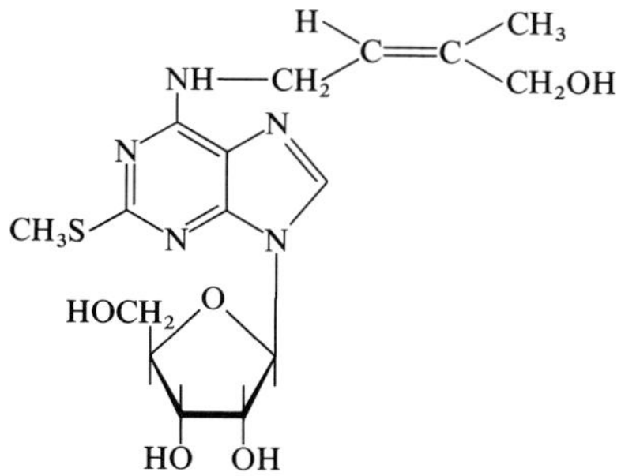

N^6-(4-Hydroxy-3-methylbut-2-enyl)-2-methylthioadenosine

isolated and identified. The first of these was N^6-(Δ^2-isopentenyl)adenosine (Hall and Srivastava, 1968). Later N^6-(*cis*-4-hydroxy-3-methylbut-2-enyl)-adenosine (this odd nucleoside has been found in soluble RNA, although it has not yet been found in any known tRNA), N^6-(Δ^2-isopentenyl)-2-methythioadenosine, and N^6-(4-hydroxy-3-methylbut-2-enyl)-2-methylthio-adenosine were disclosed (see Bartz *et al.*, 1970).

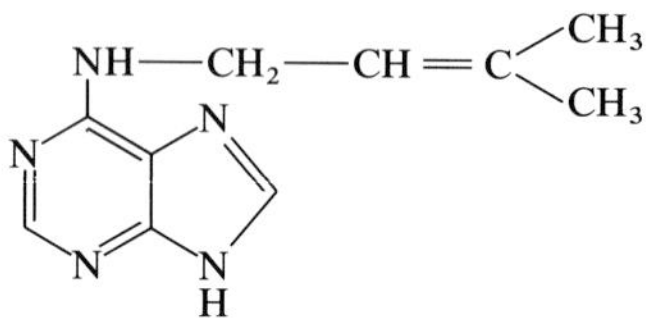

N^6-(Δ^2-Isopentenyl)adenine

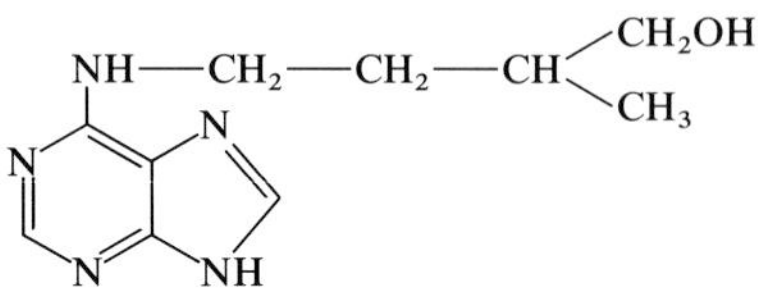

Dihydrozeatin

Moreover, the free base N^6-(Δ^2-isopentenyl)adenine has been isolated from *Corynebacterium fascians* and *Agrobacterium tumefaciens*. The trans isomer of N^6-(*cis*-4-hydroxy-3-methylbut-2-enyl)adenine (zeatin) has also been found free in higher plants (Letham, 1967a). Its ribonucleoside has been isolated from the soluble RNA of sweet corn, spinach, and garden peas. Zeatin and its riboside [N^6-(*trans*-4-hydroxy-3-methyl)but-2-enyl]-adenosine have also been isolated from the puffball fungus *Rhizopogon roseolus* and dihydrozeatin [N^6-(4-hydroxy-3-methylbutyl)adenine] has been isolated from immature seeds of yellow lupin (*Lupinus luteus*) (see Hall, 1970, and references cited therein).

The honey locust tree, *Gleditsia triacanthos*, has yielded 3-(Δ^2-isopent-enyl)adenine, or triacanthine. Triacanthine, however, has no cytokinin activity, but this appears when it is autoclaved. Because, the content of this compound (traditionally referred to as an alkaloid) varies with age and development of the plant, triacanthine has been seen as a possible storage form of the active cytokinin. During activation, a complicated rearrangement, involving the transfer of the side chain from N-3 to N-6 occurs (see Skoog and Leonard, 1968, Fig. 11).

Since N^6-(Δ^2-isopentenyl)adenosine can be enzymatically degraded in cell-free systems to N^6-(3-methyl-3-hydroxybutyl)adenine, inosine, or hyp-oxanthine (with a number of as yet unidentified intermediates) (see Hall, 1970, and Fig. 8-5), it seems that isopentenyladenosine or similar compounds are not active themselves as cytokinins, but may be converted to their active forms *in situ*. Moreover, it appears that all the modifications of the tRNA residues, i.e., the "odd" nucleosides, occur *after* the tRNA covalent chain has been transcribed from DNA (see Kline *et al.*, 1969; Yoshikami, 1970, and references cited therein). This may account for the

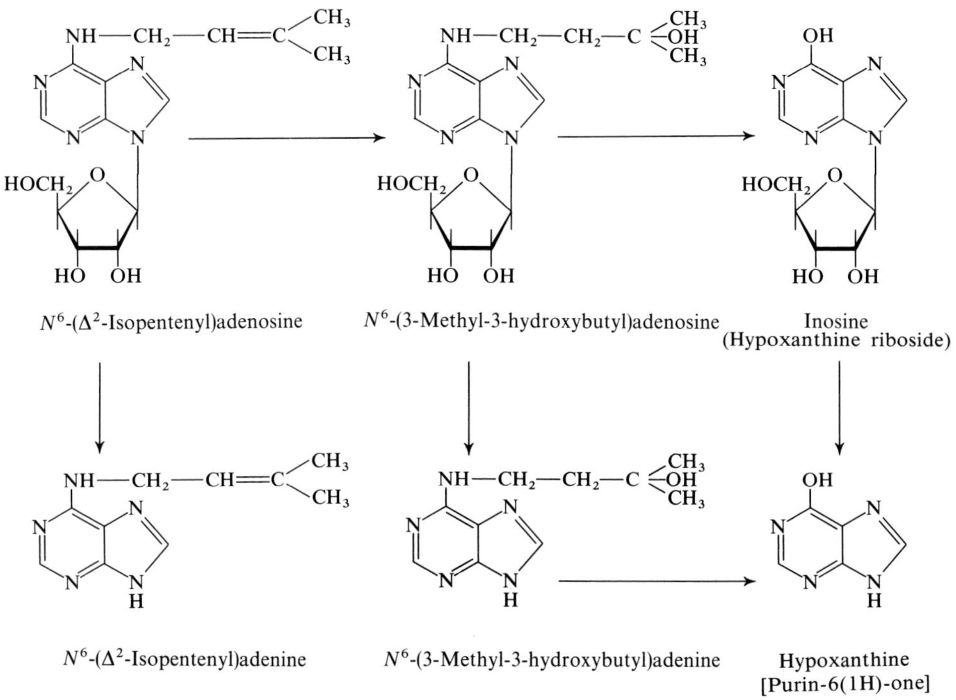

N^6-(Δ^2-Isopentenyl)adenosine → N^6-(3-Methyl-3-hydroxybutyl)adenosine → Inosine (Hypoxanthine riboside)

N^6-(Δ^2-Isopentenyl)adenine N^6-(3-Methyl-3-hydroxybutyl)adenine → Hypoxanthine [Purin-6(1H)-one]

Fig. 8-5. Suggested pathways for the metabolism of N^6-(Δ^2-isopentenyl)adenosine and related substances. [After Hall (1970).]

finding that purines and pyrimidines are often stimulatory when they supplement tissue culture media. N^6-(Δ^2-isopentenyl)adenosine, even though present in a very small amount, is regarded as very important especially because, in its own right, it has cytokinin activity. However, direct evidence to link the occurrence of this nucleoside when in tRNA to the cytokinin activity is lacking (Hall, 1970). The results obtained with this compound in relation to higher plants neither prove its cytokinin effect to be a natural one nor that the compound is a natural regulatory growth substance. Nevertheless, the observations made are very suggestive (Hall, 1970).

A key feature is that all those tRNA's that contain the Δ^2-isopentenyl group respond to codons with the first letter U (see Armstrong *et al.*, 1969, and Table 8-2). That is to say, the unusual nucleotide is localized only in transfer ribonucleic acids that accept leucine, tyrosine, cysteine, serine, and tryptophan. However, not all the "subspecies" of tRNA for leucine and serine contain N^6-(Δ^2-isopentenyl)adenosine (Peterkofsky and Jesensky, 1969). Moreover, all these tRNA's do not show cytokinin activity in the tobacco assay, even as *Escherichia coli* yields both cytokinin active and inactive serine–tRNA (Skoog and Armstrong, 1970).

TABLE 8-2
THE GENETIC CODE[a]

First letter	Second letter: U	C	A	G	Third letter
U	UUU Phe	UCU Ser	UAU Tyr	UGU Cys	U
	UUC Phe	UCC Ser	UAC Tyr	UGC Cys	C
	UUA Leu	UCA Ser	UAA OCHRE	UGA ?[b]	A
	UUG Leu	UCG Ser	UAG AMBER	UGG Tryp	G
C	CUU Leu	CCU Pro	CAU His	CGU Arg	U
	CUC Leu	CCC Pro	CAC His	CGC Arg	C
	CUA Leu	CCA Pro	CAA GluN	CGA Arg	A
	CUG Leu	CCG Pro	CAG GluN	CGG Arg	G
A	AUU Ile	ACU Thr	AAU AspN	AGU Ser	U
	AUC Ile	ACC Thr	AAC AspN	AGC Ser	C
	AUA Ile	ACA Thr	AAA Lys	AGA Arg	A
	AUG Met	ACG Thr	AAG Lys	AGG Arg	G
G	GUU Val	GCU Ala	GAU Asp	GGU Gly	U
	GUC Val	GCC Ala	GAC Asp	GGC Gly	C
	GUA Val	GCA Ala	GAA Glu	GGA Gly	A
	GUG Val	GCG Ala	GAG Glu	GGG Gly	G

[a] Cf. Sadgopal (1968) and references cited therein.
[b] ? = "Nonsense."

INVOLVEMENT OF THE ISOPENTENYL MOIETY

The suggestion has been made (Hall, 1970) that the double bond of the Δ^2-isopentenyl group may facilitate the association–dissociation process during the amino acid transfer and translocation steps of peptide bond formation (see Fig. 8-3). It is now generally granted that the modified nucleosides do confer special properties which are critical to the biological function of the nucleic acids.

Hall and his co-workers (Hall, 1970) have extensively tested the concept that the structure of the isopentenyl side chain will determine the cytokinin activity. The experiments were developed as follows.

Unfractionated yeast tRNA was modified with respect to the isopentenyl moiety of serine tRNA; this was done by treatment with aqueous iodine solutions under mild conditions. The idea here was to immobilize the Δ^2-isopentenyl groups *without* affecting the ability of the tRNA to accept serine. However, the binding of the seryl tRNA to ribosomes in the presence of poly UC was greatly reduced. Hence, the importance of the isopentenyl residue in the function of tRNA in protein synthesis remains uncertain but the change in its ribosome affinity upon alteration of the Δ^2-isopentenyl group suggests that it may affect the speed of translation by affecting the conformation of the anticodon "loop" (Fig. 8-6). Berridge *et al.*, 1970, have shown that synthetic cytokinins such as kinetin and 6-benzylaminopurine are able to bind to plant ribosomes and that adenine and its derivatives which show no cytokinin activity show much less affinity for ribosomes. Their evidence suggests binding to 83 S ribosomes and the positive correlation between the extent of binding, and the biological effect of various cytokinin analogs emphasizes the possible role of cytokinins in the control of protein synthesis.

Hall (1970) has speculated that there may be two basic mechanisms by which tRNA could play a regulatory role. In the first place the Δ^2-isopentenyl group could help to maintain the integrity of the complementary interaction between the codon and anticodon. On this view the "hypermodified" nucleoside could distort the codon–anticodon interaction in such a way that abnormal codon recognition might occur. Some of the newly discovered minor bases can cause a codon reading pattern different from those predicted by the "wobble" hypothesis. Moreover, recent evidence states that depending on the state of modification of a particular nucleoside in the anticodon, a given tRNA may read one, two, or three different code words (see Bock, 1970 for details of these findings) and Fig. 8-6.

Alternatively, certain tRNA's could be specific to certain protein molecules to be synthesized. The availability of such tRNA's would then predetermine whether that protein would be synthesized. It would be convenient, therefore, if the cell had a rapid mechanism for activating and deactivating such protein-specific tRNA molecules. To this end, the Δ^2-isopentenyl side chain possesses such reactivity and ready accessibility that simple modification of its structure might alter the codon–anticodon interaction. Also such modifications in the anticodon region do not prevent the essential attachment of the tRNA to its amino acid (Hall, 1970; Yoshikami, 1970). The tRNA–amino acid complex may remain inactive until such time as the Δ^2-isopentenyl side chain reacts and is so modified that the transfer step can occur.

From the above, the total number of tRNA molecules to which physiological activity could be traced may become very large indeed. By definition tRNA's are specific for each protein amino acid; if they become also specific for the protein to be synthesized, the potential number is virtually unlimited, and if again they become organ- or species-specific, the possibilities are even more formidable.

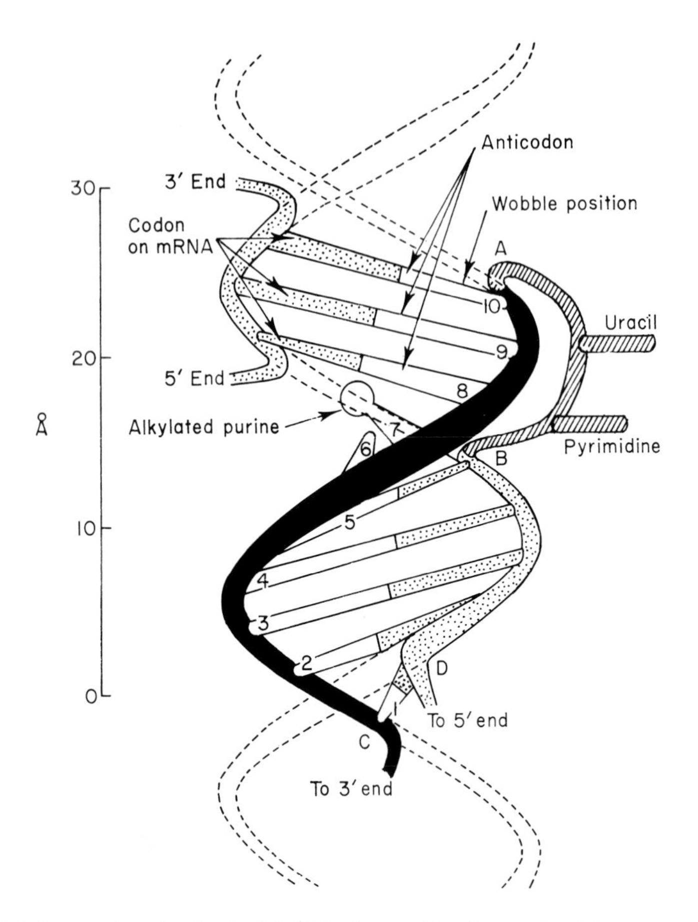

Fig. 8-6. Proposed molecular model of the base-pairing interaction between a messenger RNA codon and the transfer RNA anticodon. A, B, C, and D identify base pairs. C, D denotes the first base pair in the double helical region of the "anticodon arm," and all the bases between A and C are "stacked on one another" and follow a regular helix. The companion set between B and D and the set of three bases in the codon follow the complementary helix. In space, A and B are quite close together.

The bases in the nucleotides are shown numbered 1–10, and they are "stacked" upon one another and follow the regular helix which is shown black. The chain of the "anticodon double helix" between D and B is lightly shaded, like the "codon," to indicate that they follow the same helix, which is complementary to the black one. The two nucleotides not in the standard formation (containing uracil and pyrimidine) are represented by dark line shading, though the representation of their conformation is schematic, as they lie behind nucleotides 8, 9, and 10 in the black chain.

It is pointed out that the model suggested above provides for sufficient precision in base pairing to account for the codon–anticodon specificity which the theory requires; in addition, however, it provides for sufficient flexibility to interpret what has been termed "wobble" by Crick, i.e., a limited degree of alternative pairing in the third or "wobble" position.

Moreover, minor changes in the tRNA molecule, caused by the presence of unusual bases next to the anticodon, such as the N^6-Δ^2-isopentenyl moiety, may convey its special biological activity as a cytokinin [see text for further details (From Fuller and Hodgson, 1967)].

THE ADENYL CYTOKININS IN PERSPECTIVE

Attractive as all this may seem, many problems still remain. Only certain soluble or tRNA's contain bases with cytokinin activity, and the role of exogenous cytokinins in higher plants depends on whether the substance can be incorporated into the tRNA. More recent investigations, based on masking the 9-position of benzylaminopurine in the form of the compound 6-benzylamino-9-methylpurine (benzyl-7-^{14}C), calls into question the interpretation made by Fox (1966) (see page 131). This synthetic cytokinin was not incorporated into any RNA fraction whatsoever but, nevertheless, it had activity in the soybean callus assay comparable to the "unmasked" compound (Kende and Tavares, 1968). Moreover, although *E. coli* contains cytokinins in its soluble RNA, no cytokinin auxotrophic mutants have been found (contrary to expectations) using standard microbial genetic procedures. Statistical procedures indicated that the probability for the existence of such cytokinin-dependent mutants is extremely low indeed (Kende and Tavares, 1968).

The conclusion drawn from this work is that the "incorporation of cytokinins such as benzyladenine into RNA is not related to the action of this hormone" (Kende and Tavares, 1968). Since this conclusion dismisses the very foundations of the current hypotheses of cytokinin action, it has not gone unanswered. Skoog and Armstrong (1970) have pointed to the possibility that the methyl group on the 9-position could be both lost and metabolized and there is some evidence for this (Skoog and Armstrong, 1970). The inability to find cytokinin-requiring auxotrophic mutants of *E. coli* has been merely dismissed as negative evidence. The evidence suggests that the cytokinins may represent metabolic or breakdown products of the nucleic acids (Hall, 1970). Therefore, the absolute dependence of cells on their exogenous supply of cytokinins seems difficult to understand.

Hall *et al.* (1970) have shown that strains of cultured tobacco tissue, which require adenyl cytokinin to grow, contain N^6-(Δ^2-isopentenyl)-adenosine. This result has stimulated these workers to state that "the occurrence of this cytokinin in the tRNA of many organisms raises the question of whether its occurrence in tRNA is related to the mechanism of action of cytokinins." Moreover, mevalonic acid is a precursor of N^6-(Δ^2-isopentenyl)adenosine in these external cytokinin-requiring strains and the biosynthetic pathway does not appear to be defective. Finally, cultured tobacco pith tissue has been shown to have enzyme systems that can metabolize N^6-(Δ^2-isopentenyl)adenosine (see page 137). Guern (1970) has also shown that adenosine aminohydrolase catalyzes the *in vitro* hydrolysis of various N^6-substituted adenosines to yield inosine, and has evidence that enzymes of the adenosine deaminase type control the level of cytokinins within the cell. This means that any N^6-(Δ^2-isopentenyl)adenosine made available from tRNA breakdown can be metabolized. In these adenyl cytokinin-requiring systems, at least, it seems that the requirement for external cytokinin rests not with the inability to synthesize N^6-(Δ^2-isopentenyl)adenosine but with its subsequent utilization. In this connection it is interesting to speculate on the similarity of hypoxanthine which results from the metabolic breakdown of N^6-(Δ^2-isopentenyl)adenosine (see page 137) and the chromophore of the cell division factor isolated from crown gall tumor cells (Wood, 1970) as being more than a coincidence.

Cytokinins applied to moss protonemata (see Table 4-2) cause bud formation, but when the cytokinin is removed by extensive washing, the buds revert to filaments (Brandes and Kende, 1968). This indicates that the cytokinin does not act only as a "trigger" but must be present at least for a "critical period" until bud differentiation is stabilized. In addition, it has been shown, by the technique of radioautography, that benzyladenine is selectively accumulated by the cells that respond to cytokinin; this suggests that there are binding sites that determine whether cells do or do not respond (Brandes and Kende, 1968).

Still another anomaly, raised by Galston and Davies (1969, 1970), is that the free base zeatin as isolated and identified from corn, and more recently from fungi, is in the trans configuration, whereas hydrolyzates of tRNA from the same material yield the cis isomer. Cis–trans isomerizations are well known, however, (see Mayo and Walling, 1940). [The configurations of the side chains in ribosylzeatin and 2-methylthioribosylzeatin from wheat germ tRNA are not yet known (Burrows *et al.*, 1970).] Nevertheless, it has been shown that side-chain planarity is important in conferring high adenyl cytokinin activity (Hecht *et al.*, 1970b). *trans*-Zeatin riboside seems to be much more active than the cis isomer. The addition of substituents to the double bond of the side chain results in lowering cytokinin activity (Hecht *et al.*, 1970b).

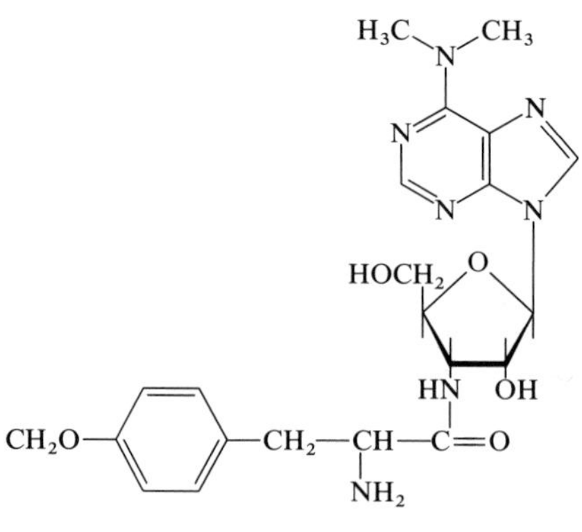

Puromycin

Experiments with animal systems have had the objective of testing whether they yield similar results to those obtained with plants. Most of the experiments have been performed with synthetic compounds like kinetin and "the results have generally been negative," although some work done with N^6-(Δ^2-isopentenyl)adenosine in various animal tissue culture lines has shown that it is a potent inhibitor. It is well known that many adenosine derivatives such as the antibiotic puromycin are inhibitory to the growth of plant, animal, and bacterial cells (Fox *et al.*, 1966). It is beyond the scope of the present work to analyze the data on animal cells. Suffice it to say that the mode of action of N^6-(Δ^2-isopentenyl)adenosine in animal systems, as indeed in plants, is still unknown (Hall, 1970; Skoog and Armstrong, 1970).

The extensive and obviously careful work summarized above has been directed toward involving the cytokinins which stimulate cell division in higher plants in some limiting aspect of protein synthesis. To the extent that *any* substance or mechanism that stimulates cell division must also permit or stimulate the concomitant synthesis of protein this is an understandable objective. The work described, however, leans heavily on current views of protein synthesis (Allende, 1969; Boulter, 1970) which derive from the bacterial, virus, and animal (rat liver) systems (see Lipmann, 1969; "Mechanisms of Protein Synthesis," 1969). Accepting the fact that the "adenyl cytokinins" do possess the properties attributable to them, one would still be at a loss to explain, in similar terms, the many other substances (not adenines) that undoubtedly also stimulate cell division. The curious, but unquestioned role of IAA, which is essential to the activity of the adenyl compounds still does not have any explanation comparable to that advanced for the role of the adenyl compounds per se. The role of inhibitors also merits comment, for they must intervene, often reversibly, to regulate cell division and cell enlargement as, e.g., in the opposed actions of gibberellin, cytokinins, and abscisic acid (Addicott and Lyon, 1969; Milborrow, 1970). It is curious that the action of inhibitors has not been as explicitly explained as has that of the cytokinins.

Since the experiments which invoke cytokinins in the tRNA's have utilized, essentially, the same systems that have been involved in the experiments on the genetic code (see Sadgopal, 1968, and references cited therein; Watson, 1970), it seems surprising that the cytokinins for higher plants have been neither easily assayable using bacteria nor have they appeared as dramatically stimulatory to the growth of microorganisms. The workers who have carried out the experiments with labeled synthetic cytokinins have stressed, repeatedly, the extensive metabolism which these substrates undergo, so much so that over 95% of the material supplied is dispersed in other ways (Fox and Chen, 1967).

Fox (1970) has recently emphasized that "cytokinins are probably not active as such but are metabolically transformed to compounds having biological activity." This new viewpoint has developed as a result of work carried out on the metabolism of "urea cytokinins," such as 1,3-diphenylurea (see page 104). Cultured tissues of soybean and tobacco were "starved" for cytokinins and then exposed to labeled 1,3-diphenylurea (^{14}C and/or ^{3}H). In soybean tissues the 1,3-diphenylurea was not metabolized; however, in tobacco, the compound was metabolized to other substances. The key point here is that in the adapted tobacco strains, 1,3-diphenylurea is active as a cytokinin because it can be metabolized. The strains of soybean tissue thus far available are only able to utilize adenyl cytokinins. Even in these adenyl cytokinin-requiring strains, free base when supplied as labeled 6-benzylaminopurine exists only for a short time and is converted to a number of products, such as the 5′-monophosphate of benzyladenosine. Within a matter of hours, however, the ribonucleotide disappears and a new, long-term, as yet, unidentified, metabolite appears. This substance when reextracted behaves on examination like a nucleotide with high cytokinin activity. Therefore, results of this sort tend to provide further evidence that these substances are not active, as such, but the tissues which can utilize them convert them to an active form (Fox, 1970).

N-(Purin-6-ylcarbamoyl)threonine riboside [*N*-(nebularin-6-ylcarbamoyl)-threonine] has been detected in a tRNA of an autonomous strain of tobacco which was fed labeled adenine, but when this compound and its corresponding base were tested (in the presence of an auxin) upon a cytokinin-dependent strain of soybean there was no activity (see Dyson *et al.*, 1970). These investigators attributed the lack of activity to poor absorption, or, alternatively, to the fact that the *N*-(purin-6-ylcarbamoyl)threonine riboside must be metabolized in order to be rendered active. Synthetic analogs, on the other hand, showed growth promotion. *N*-Purin-6-yl-*N*′-allylurea and *N*-purin-6-yl-*N*′-isopentenylurea were two of the active analogs synthesized. Dyson *et al.* (1970) note that the "phenylurea derivatives have growth-promoting properties by virtue of their structural similarity to the naturally occurring purinylurea nucleoside" [i.e., *N*-(purin-6-ylcarbamyl)-threonine riboside].

The dilemma, therefore, is twofold. First, can we be sure that the cytokinin effect is to be attributed solely to that part of its metabolism that involves its entry into tRNA? The presence of the "odd" bases with cytokinin activity may in this context be a coincidence. Second, would one expect that a substance which is part of so sensitive a regulatory mechanism involving so critical a feature as the control of cell division would be dispersed throughout the cell's metabolism? Surely this would be an effective means to ensure their relative inactivity.

Laudable as this overall endeavor has undoubtedly been, and still is, it depends upon the need to synthesize highly specific proteins which are necessary for angiosperm cell divisions and any other cytokinin-mediated activity. This requires that the normally amino acid-specific tRNA's become also specific for the various proteins to which they relate; and the whole approach is linked to the applicability of the gene–tRNA–mRNA–ribosomal template hypothesis of protein (see Cox, 1968) synthesis which, in higher plants, still lacks the definitive proof at keypoints (Allende, 1969; Boulter, 1970), e.g., the existence of the appropriate mRNA's, which is comparable to that obtained with bacterial and viral systems.

The views on the nature and mode of action of the adenyl cytokinins have leaned so heavily on the conventional molecular biological interpretation of protein synthesis (see page 130) that it is not possible to deal with one problem without assessing the status of the other. If the adenyl cytokinins act by their incorporation into tRNA's, and hence determine the formation of specific protein, and if such RNA's in cytoplasm are longer,

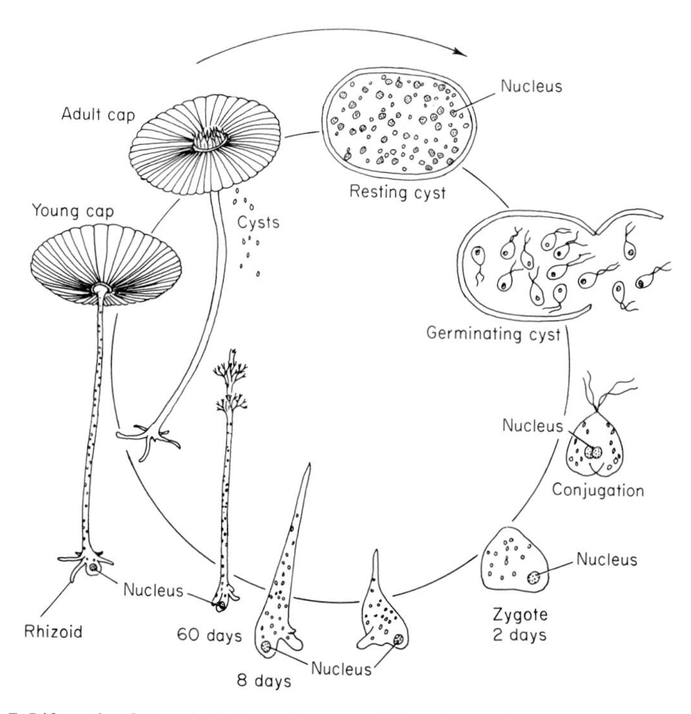

Fig. 8-7. Life cycle of *Acetabularia mediterranea.* When the cap forms, the large basal nucleus gives rise to many nuclei which spread throughout the alga, including the cap, which produces resistant forms (called cysts). As the cysts germinate, motile gametes are formed and escape. After union of the gametes and the formation of a zygote, the cycle is repeated (Brachet, 1957).

rather than shorter lived, then the degree of regulation by such substances would seemingly not be a sensitive one. Harris (1968) has reinterpreted the work on *Acetabularia*, the organism that has become familiar because, through much of its life cycle, it has a characteristic single nucleus which gets larger without division and which can be removed (see Fig. 8-7). In this system, the cytoplasm continues to operate in protein synthesis *without* the nucleus (Brachet, 1968). Therefore, the persistent cytoplasmic templates do not require to be as continually activated via the nucleus and mRNA's as the conventional views require because both synthesis and morphogenesis persist in the enucleated state. [In this system auxin, IAA, and its antagonist, triiodobenzoic acid (TIBA) have been observed to promote and antagonize elongation and cap formation of enucleated plants (Thimann and Beth, 1959).] Other enucleated cells like *Stentor* and *Spirogyra* also enlarge and synthesize for long periods in the absence of the nucleus and observations on enucleated frog oocytes have shown that they still respond to progesterone by further maturation (cited from Jaffe, 1969). Thus, Harris concludes that "the time at which a particular gene is transcribed into RNA has no immediate connection with the time at which this RNA is translated into protein. The templates for protein synthesis pass to the cytoplasm in a form which is, on the whole, resistant to intracellular degradation. Initiation, regulation, and suppression of protein synthesis on these essentially stable templates is effected by cytoplasmic mechanisms which can operate perfectly well in the absence of the nucleus." In other words, these "cytoplasmic mechanisms" could be accessible to regulatory control even without the necessarily continuous involvement of the nucleus or the intervention of the cytokinin hormone at the level of tRNA along the conventional chain (see page 130).

The genetic operator model (see page 123) requires that the templates for protein synthesis be short-lived; otherwise "switching-off" the appropriate gene would not "turn off" the production of the corresponding protein. This seems not to be so in eukaryotic organisms. Despite the intensive search for transient "messengers," mRNA's have never been directly demonstrated in a higher plant system, although they have been amply demonstrated, and the nucleotide sequence determined, for bacterial virus coat protein messenger (Nichols, 1970; Steitz, 1969).

The general evidence for mRNA flows from the use of synthetic polynucleotides applied to isolated ribosomes and their ability to function there as a template for the synthesis of peptides for which they code. Even so this does not prove that ribosomes *in situ* are similarly coded, for to quote Harris, "It could be argued that the added polynucleotide masks the normal template surface of the ribosome and that the synthesis of the polypeptide takes place on this artificial mask." To demonstrate a natural messenger adequately, one should do this in relation to the synthesis of a highly specific protein. Laycock and Hunt (1969) have elegantly shown this convincingly. A cell-free system from *E. coli* primed with an RNA preparation from rabbit reticulocytes produces a material that possesses the characteristics of rabbit globin. The fact that polypeptide chains can be made which behave in all respects like rabbit globin using messenger RNA from reticulocytes but transfer RNA and all other components from *E. coli* is strong evidence for the universality of the code. Despite this work, however, no such evidence is available yet for plant systems, and until this is done one is forced to face facts.

In the outcome, therefore, can we say how the genetic information is controlled in higher plants even for the particular phenomenon of protein synthesis? Even prior to the Jacob–Monod hypothesis, McClintock located nongenic control centers in chromosomes which modify the expression of genes (McClintock, 1951, 1967). There are abundant physiological reasons why cytoplasmic regulation should also be considered. But one should not restrict the biochemical control of growth to protein and enzyme synthesis alone. To divide and grow, and divide again, cells perform many other functions; of these, internal secretions of water and solutes, organic and inorganic, are important, even indispensable, steps in the cell cycle and in their ontogeny.

These events involve membrane phenomena and energy relations which may involve proteins as enzymes, but in more than merely a synthetic role. In other words, to perform these complex events, cells present in addition to the possibility of hormonal control, many other features than those which are comprehended in the schemes outlined (see Figs. 8-1 and 8-3). It is even pardonable to read some recognition of this fact into Skoog's cautious summing up in the words, "The identification of cytokinins as constituents of tRNA has not resulted in immediate clarification of the mechanism of cytokinin action." Moreover, the range of the "central dogma" needs now to be reexamined because of the finding that although DNA normally directs the formation of RNA, i.e., the transcription process, it is now conceded that, conversely, RNA can serve in some systems as a template for the synthesis of DNA; this reverses, therefore, the direction of transcription (Baltimore, 1970; Temin and Mizutani, 1970).

The only final conclusion that can be drawn is that growth in the cells of higher plants, whether these grow in isolation or *in situ*, is a highly complex event about which a great deal remains to be learned. Often empirical observations concerning chemical controls (exogenous and endogenous) may be made, but until the sites and modes of action can be precisely determined, opinions should not be so polarized toward preconceived or simplistic explanations so that the course of other independent inquiries tends to be stifled.

CHAPTER 9 Concepts and Interpretations of Growth Regulation

"Plant Hormones" and "Action at a Distance"

From the outset, plant growth regulators, interpreted as hormones, have been heavily involved with "action at a distance." Their ability to move within the plant body has been a paramount consideration. The long established, predominantly basipetal, movement of indole-3-acetic acid needs no further comment here (Goldsmith, 1968, 1969; Pilet, 1961; Vardar, 1968). Whenever there is a pronounced site for perception of a stimulus and a definite location for the response, a means of transfer needs to be invoked. Moreover, the innumerable synthetic growth regulators and herbicides must rely on ease of absorption where applied, e.g., often via the foliage, and rapidity of movement to their sites of action (McCready, 1966; Moreland, 1967). In this connection, the specificity of auxin transport has been examined by Hertel *et al.,* (1969). These investigators found that in contrast to the basipetal, polar transport of the auxins 2,4-D and IAA, no such movement was found for benzoic acid or for gibberellin A_1. A comparison of the α- and β-forms of NAA showed that the α-isomer which is active as a growth regulator is transported, but the inactive β-isomer is not. Moreover, the (+) and (−) isomers of 3-indole-2-methyl-acetic acid move basipetally, as would be expected of auxins, but the (+) form which is more active biologically was transported more readily. Antiauxins such as *p*-chlorophenoxyisobutyric acid inhibited transport of auxin and its analogs and thereby any auxin effects. This evidence has suggested to Hertel *et al.* (1969) that the "recognition sites" for primary auxin activity are intimately associated with, or may be identical with, the sites for auxin transport.

Whereas it is a characteristic of the naturally occurring or synthetic auxins that they are readily mobile within the plant body, the converse is true of the adenyl cytokinins. Experiments utilizing radioisotopes directed attention (Mothes, 1964) to the localized stimulus of kinetin which caused movement of ^{14}C-labeled amino acids from another part of a leaf lamina to the area where the kinetin was applied and the response (protein synthesis) occurred (see Fig. 9-1). In the principal assays, the cytokinin is usually in direct contact with the cells affected.

Long established morphogenetic ideas attribute a formative role in the growth of the shoot to substances elaborated in roots, and this is over and above such metabolites as may arise, nutritionally, in the root from carbohydrates via the shoot and reduced nitrogen compounds via the root. In Chibnall's book on "Protein Metabolism in the Plant" the chapter on "Regulation" concluded with the query whether hormone-like substances from the root could control protein metabolism in leaves and, thus, the behavior of shoots (Chibnall, 1939). Chibnall was struck by the dramatic consequences for their metabolism which ensue when leaves are detached from contact with roots. Hence, regulatory substances emanating from roots would need to move readily in the acropetal direction, whereas normally applied growth regulatory substances are furnished to the shoot, so that it is the basipetal movements that are most conspicuous (McCready, 1966).

Indirect indications suggest the normal upward movement of regulating substances (Carr and Reid, 1968). Minute fragments of shoot apices are proverbially difficult to culture (Nougarède, 1967), but if root primordia develop the consequences for growth of shoots is dramatic. Even though embryos may store much total nutrient in their cotyledons, it is often the radicle which first becomes active and emerges. Therefore, it is not surprising that there are some recent indications that the adenyl cytokinins can move acropetally in ways not envisaged in the first experiments of Engelbrecht and Mothes on isolated leaf laminae (Mothes, 1964). Vaadia and his coworkers (Itai *et al.*, 1968) attribute the effects on shoots of salinity in the root environment to cytokinins which are transferred to the shoot. But the rates of acropetal movement observed are not comparable to the highest speeds of translocation of labeled carbon compounds (Nelson, 1963). The principal effects attributable to gibberellins, whether they arise from the parasitic fungus or are due to low temperature stimulus of plants during vernalization, could be at the site of gibberellin formation. Moreover, although the stimulus to flower formation is obviously mobile, there seems to be no evidence that the mobile entity is a gibberellin, even though the

Fig. 9-1. Effect of applied kinetin on the distribution of ^{14}C from glycine applied locally to the right-hand member of paired leaflets of *Vicia faba*. The kinetin was variously applied as follows. A, No kinetin applied; B, three applications of kinetin to the left-hand leaf while still attached—the paired leaflets were excised and the ^{14}C-glycine subsequently added; C, the right-hand leaflet when attached received kinetin and later, when excised, the ^{14}C-glycine; D, the excised leaflets were rooted, the kinetin later applied to the left leaflet and the ^{14}C-glycine to the right. In A, the ^{14}C from glycine spreads through the leaflet which received it and slowly into the left; in B, the transfer to the left was greatly enhanced by the kinetin it received; in C, the radioactivity is restricted to the kinetin-treated leaflet; in D, the kinetin effect to the left-hand treated leaflet was overcome by the growth substances which moved from the roots to the right-hand leaflet. (Photograph supplied by Dr. H. K. Mothes, Halle/Salle, German Democratic Republic.)

gibberellins may be a part of the response elicited (see Lang, 1965, 1966). The very fact that the claims for gibberellin movement in the plant body seem to rest on its passive flow would tend to rob the movement per se of a major part in the mechanism of regulation.

The paradox is that the hormone concept in plants, built around auxin, began with "action at a distance" and early stressed the importance of the movement of stimuli carried by "hormones." As our knowledge of the chemistry of regulatory substances has proliferated, we seem now to know very little about their movement; certainly not enough to invoke the means of their movement as a limiting factor in their modes of action.

There is relatively little mutual usage of the same growth factors by plants and animals. Few of the specific physiologically "active-at-a-distance" substances of the animal body, e.g. hormones, have any direct effects on plants. [An example of an indirect effect is that of human chorionic-gonadotropin (HCG) steroid hormone on plants (see Leshem *et al.,* 1969). The view here is that the gonadotropin stimulates conversion of mevalonic acid to steroids and, thus, the availability of precursors for gibberellin biosynthesis is diminished. Moreover, auxin levels are decreased by HCG, possibly by lowering the IAA peroxidase levels (Leshem, 1970). Bonner *et al.* (1963) have shown that inhibitors of steroid biosynthesis suppress floral induction of short-day plants.] On the other hand, plants produce many compounds which do have extreme physiological effects on animals (as the great array of drugs from plants), though not on the plants in which they originate. Even carcinogens of plant origin are not conspicuous as cell division factors in plants (Deysson, 1968; Fishbein *et al.,* 1970; Kihlman, 1966). Many other substances known to have physiological action on plants, e.g., colchicine, may not produce their effect on the plant that synthesizes them. This could be due to compartmentalization. Another example is the compound fluoroacetate (CH_2FCOOH) that inhibits the citric acid cycle by giving rise to fluorocitrate. Fluorocitrate is formed by the enzyme-catalyzed condensation of fluoroacetyl-CoA and oxalacetate, and competetively inhibits aconitase. *Dichapetalum cymosum,* a very poisonous plant from South Africa, contains large amounts of fluoroacetate, but seems to be unaffected by the presence of the compound (Peters, 1954; Peters *et al.,* 1960).

The first class of plant growth regulators, i.e., the indoles, are related to animal waste products (Stowe, 1959). Conversely, newer developments now show that many powerful substances that act as insect hormones may be of plant origin and this raises many interesting questions of insect–plant interrelationships (Heftmann, 1970; Staal, 1967; Williams, 1969). Many antibiotics are by-products of plant (e.g., fungal) metabolism (Brian, 1957). These and other substances which act as plant pathogens or naturally occurring antifungal substances are not dealt with here (Dimond, 1959; Fawcett and Spencer, 1969; Owens, 1969).

Presentation to the Sites of Action

The role of the size and shape of a molecule within a given class of growth regulators, as a determinant of its action by virtue of its ability to penetrate to the appropriate site, has been emphasized. Properties such as lipid solubility, which is modified by substituent groups and by side chains, are relevant here.

In the ω-substituted phenoxyalkanecarboxylic acids, their activity is determined by the length of a side chain. Compounds with side chains with an *odd* number of carbons may be degraded enzymatically to the corresponding phenoxyacetic acid by β-oxidation, whereas the β-oxidation of side chains with *even* numbers of carbons does not result in an active compound, i.e., phenol is formed (see Fawcett *et al.,* 1956, and Fig. 9-2). This effect appeared in several kinds of assay, and it was also shown to hold in the synergism between coconut milk and the substituted phenoxy acids in

Fig. 9-2. Selective herbicidal activity in relation to chemical structure. Upper row, charlock (*Sinapis arvensis* L.) plants sprayed with the substances specified; lower row, clover (*Trifolium* spp.) plants sprayed with the same chemicals. Left to right: A, 2-methyl(4-chlorophenoxy)acetic acid; P, propionic; B, butyric; V, valeric; C, caproic; H, heptanoic; Con, control (untreated). In charlock, the butyric and caproic derivatives are as active as the acetic acid compound by virtue of the β-oxidation. In clover, only the acetic compound is active, since the β-oxidation enzymes are lacking. (Photograph supplied by Dr. R. L. Wain, Wye College.)

$$C_6H_5OCH_2CH_2COOH \rightarrow [C_6H_5OCOCH_2COOH] \rightarrow [C_6H_5OCOOH] \rightarrow C_6H_5OH$$
$n = 2$ Phenol

$$C_6H_5OCH_2CH_2CH_2COOH \rightarrow [C_6H_5OCH_2COCH_2COOH] \rightarrow C_6H_5OCH_2COOH$$
$n = 3$ Phenoxyacetic acid

their ability to stimulate potato tuber tissue to grow (Shantz *et al.*, 1955). Wain has also emphasized, especially in the case of α-(2-naphthoxy)-, α-(2,4-dichlorophenoxy)-, and α-(2,4,5-trichlorophenoxy)propionic acids, the often overriding importance of one active enantimorph (+) over the other (−) which usually possesses little or no activity (Wain and Fawcett, 1969). These subtleties of molecular configuration of active molecules all suggest geometric considerations which determine their ability to fit into an active site, as on an enzyme or protein surface. These ideas are summarized in a diagram reproduced from Wain and Fawcett (Fig. 9-3).

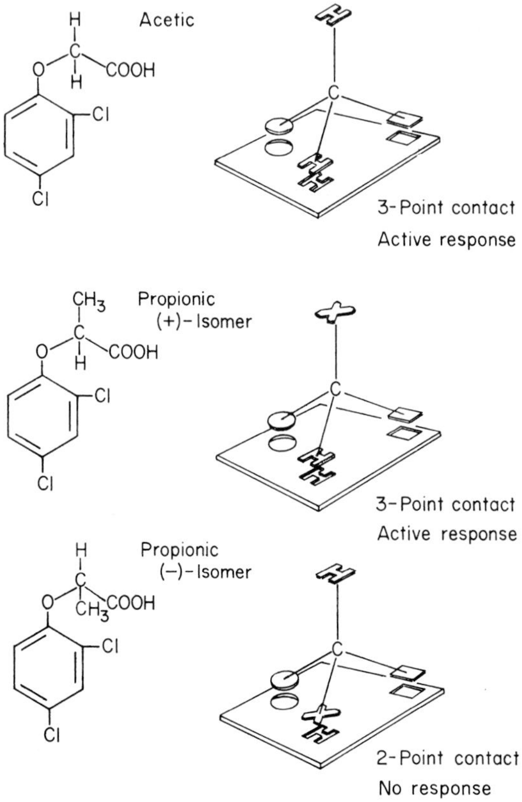

Acetic

3-Point contact
Active response

Propionic
(+)-Isomer

3-Point contact
Active response

Propionic
(−)-Isomer

2-Point contact
No response

Fig. 9-3. Diagrammatic representation of the presentation of a growth regulator to its active site. The idea illustrated is that an active aryloxy acid presents three functional groups (the unsaturated ring, the α-H atom, the COOH) which engage with three complementary "points of contact." [From Wain and Fawcett (1969).]

Wain has also made use of the enzymatic degradation of some substituted molecules to the active form in the design of selectively active growth regulators, fungicides, or herbicides by capitalizing on the distribution of enzymes in plants that need to be selectively treated (Fawcett *et al.*, 1956; Wain and Carter, 1967).

An outstanding example of the production of an active molecule from another one supplied is the case of ethylene (Cooke and Randall, 1968; Mapson, 1969; Palmer *et al.*, 1967). We have already pointed out (see page 115) that some feel that many synthetic growth regulators owe their action to their ability to give rise in the treated plants to ethylene, due to the presence or absence of the appropriate enzyme in the plants (Pratt and Goeschl, 1969). Even so, the mode of action of the ethylene so produced is still difficult to specify.

MECHANISMS OF ACTION

A growth-regulating compound could act at any point in the cell cycle of synthesis or replication. It could affect cell enlargement or enhance metabolism by affecting transfers across membranes or by activating enzymes, stimulate translocation of nutrients over long distances (i.e., to, or from, growing regions), or it could intervene at the higher levels of control where morphogenesis is determined. But at whatever level such a substance acts, there is the question of its mechanism.

There is virtually no class of plant growth-regulating compounds for which there is an unequivocal molecular explanation of biological action.

Concepts Involving Cell Walls in Growth Regulation

Many explanations have been offered for the action of IAA, the oldest plant growth regulator (see Galston and Purves, 1960; Ray, 1969; Thimann, 1969). IAA was at one time credited with an ability to affect the methylation of cell wall galacturonic acids and thereby facilitate elongation (see Steward, 1968, and references cited therein). This concept involves current ideas about what constitutes a cell wall, for this will influence interpretations of the action of growth regulators as they affect cell division and enlargement.

During the elongation growth of the coleoptile, the classical organ for IAA assay, the extension of a preformed cellulose cell wall is dramatic. At the other extreme, when cells divide, the new cell walls of daughter cells form at the cell plate. Conventional or classical concepts of the cell wall have regarded it as external to the protoplast, i.e., external to the plasmalemma, and as a nonliving secretion product of the cell. In sharp contrast to these ideas, some would now locate in the cell wall events described as "cell wall metabolism" (see Lamport, 1970). This would earlier have seemed to be a contradiction in terms. However, these views have opened the door to speculations on the metabolic action of growth-regulating substances (IAA, cytokinins, or gibberellins), in what was hitherto regarded as the inanimate cell wall.

A line of research that bears upon this question originated with the observation that tissue explants (of carrot or potato tuber), caused to grow by the factors of coconut milk, contained a combined hydroxyproline-rich moiety in what was otherwise the total protein of the cell. This had all the characteristics of a structural entity because it was not metabolized, did not "turn-over," it incorporated proline directly, and hydroxylated it subsequently in a manner reminiscent of collagen. The evident enrichment of the cultured cells in this hydroxyproline-rich moiety, in contrast to its relative abscence from the cells *in situ*, suggested that its presence was one of the consequences of the growth so induced (Pollard and Steward, 1959; Steward and Pollard, 1958).

Meanwhile, Lamport had grown cells of *Acer pseudoplatanus* in a liquid medium which also contained coconut milk and 2,4-D. From the outset Lamport assigned the combined OH–proline to the cell walls and spoke about "extensin," a wall protein deemed to be an essential part of the mechanism of cell wall growth and extension (see Lamport, 1965; Ray, 1969, and references cited therein). The principal isolates from crops of *Acer* cells, however, after alkaline hydrolysis of the insoluble residues, proved to be various glycopeptides containing hydroxyproline. This area of biochemistry is only of interest in the context of this work if auxins, cytokinins, or gibberellins function by fostering the synthesis of these substances in the walls of the living growing cells they affect.

The work on the cultured carrot system, which first drew attention to the role of combined hydroxyproline as a feature of the growth induction event, never located the substance in question in the cell wall proper (Israel *et al.*, 1968). Radioautographic studies performed with ^{3}H-proline *and* ^{14}C-proline on cultured carrot cells, and also on growing *Valonia* aplanospores (Steward *et al.*, 1970), located the labeled products in the protoplasm, rather than the wall (see Fig. 9-4). The interesting disclosure was that the entering proline accumulated first in the nucleus and nucleolus and eventually spread through the ground granular cytoplasm, up to, but not beyond, the plasmalemma. The idea emerged that this nonmetabolizable moiety is a submicroscopic structural element in the cytoplasm and, with its further purification, difficulty was encountered in separating it from nucleic acids (Steward *et al.*, 1970).

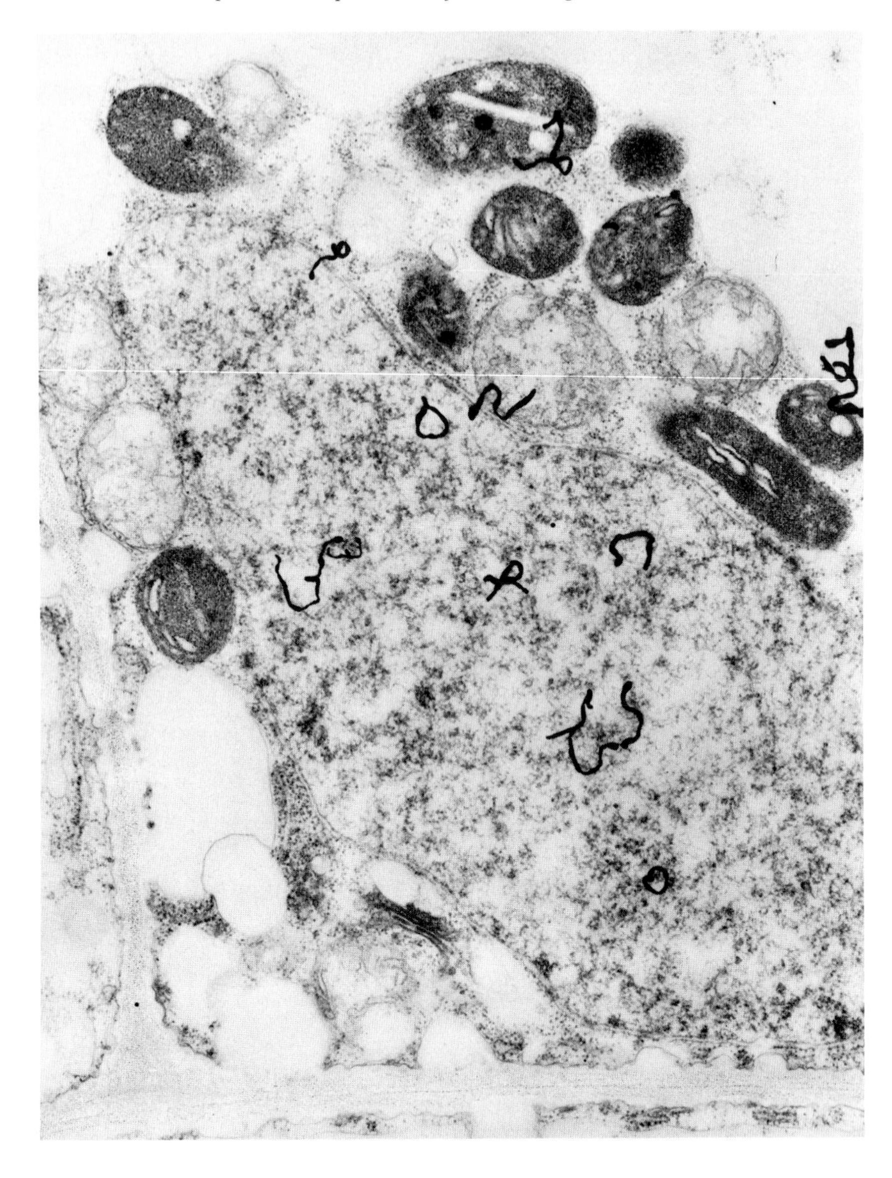

Fig. 9-4. A high resolution radioautograph showing cultured carrot cells which had been labeled with ^{14}C-proline. The figure shows thin cell walls developed during contact with ^{14}C-proline, adjacent protoplasts with organelles, including a nucleus bounded by its limiting membrane. Whereas the label is present throughout the protoplasm, especially in the nucleus, it is not evident in the cell walls (see work of Israel *et al.*, 1968). ($\times$ 26,000.)

The situation in the *Acer* cultures of Lamport seems to be as follows. The free cells grow isolated from each other; they show no ability to organize; the necessarily heavy crops of cells that are harvested for biochemical analysis then consist predominantly of dead cells. Those with abundant cytoplasm and which can still grow are infrequent. Many of the cells in such cultures have somewhat sculptured walls. When, in the laboratory of one of us (F.C.S.), living viable *Acer* cells are labeled with radioactive ^{14}C-proline, the label is present in the protoplasm just as in the case with carrot cells. The label only enters the wall effectively when it consists of the sculptured forms mentioned and these are in cells that do not then have active protoplasts. The wall sculpturing, seen clearly only with the electron microscope, is an aspect of biochemical differentiation in senescent cells; moreover, cells which have been so long cloned and cultivated may have acquired some restricted properties. The cells that are still active, when walls are still very thin, retain the great bulk of their labeled proline in their portoplasm.

To make a long story short, the position seems clear. If one uses the term "cell wall" to mean a "cytologically recognizable cell wall" in a viable living cell, then there is no need to invoke it as the principal site for *either* the combined hydroxyproline of the cells *or* the metabolic properties normally ascribed to the protoplasm. The presence of noncellulosic substances in cell walls is familiar (see Northcote, 1969). These may be fatty substances, lignins, hemicelluloses, etc., but their accumulation there in cells that are often, or usually, dead has no necessary bearing on the role of a wall during the early growth of a cell, or upon the site of action of the growth substances which affect it; it is to be expected that the latter would act on sites in the protoplasm.

Concepts Relating Auxins and Ethylene

This relationship deals with one of the strange episodes in the understanding of chemical growth regulation in plants. It involves auxin, the prototype of a "plant hormone," on the one hand, and, on the other, the gas ethylene, a very simple molecule which can act as a plant growth regulator.

Plant physiological responses to the stimulus of exogenous ethylene have a long and familiar history which stems from the early observations of effects on plants due to traces of coal gas and to the use of products of combustion, as in flu gases, to induce or hasten ripening or coloration of fruits. Direct effects of traces of ethylene on the respiration of many plant organs such as leaves, potato tubers, apples, and bananas were all familiar to the plant physiologists of 30 to 40 years ago. Until the advent of gas chromatography the epinasty test (see Fig. 3-4) was regarded as the most convenient and sensitive test for ethylene gas. But the actual metabolic production of traces of ethylene by ripening fruits, unequivocably demonstrated chemically by Gane (1934) in the case of the banana, and the ability of ethylene produced by one plant organ to trigger off a response, such as the onset of the climacteric in fruits in another, endowed this extremely simple gaseous molecule with one of the essentials of hormone action, namely, "action at a distance." The similarity of many of the responses of plants to ethylene and to auxin was fully appreciated in the early work of Zimmerman and Hitchcock (1933) and Zimmerman and Wilcoxon (1935), and the knowledge of these facts, even though it lingered long in the background, was to plague those who variously speculated on the structural characteristics of a molecule that could equip it to function as and properly be classified as an auxin or "plant hormone." But the fact, readily revealed since the advent of gas chromatography, that has awakened so much recent interest in ethylene is that, ethylene is produced as a response to auxin application.

The Ottawa conference of 1967 on plant growth regulators (see Wightman and Setterfield, 1968) contains two papers, one by Burg and Burg (1968) and the other by Goeschl and Pratt (1968), which essentially tell the story as it was then known. Since that time, the following developments have occurred, and for these summarized observations we are indebted to Dr. Stanley Burg.

Consistently, and whenever a plant responds to applied, i.e., unbalanced, applications of auxin, ethylene is produced. This occurs whether the organs in question are roots or shoots, or whether the response occurs in light or darkness. The level of applied auxin which is needed to evoke ethylene may be as low as 10^{-7} *M*, and from this level ethylene output increases with applied auxin, even up to concentrations above those at which the organs are already saturated with auxin in terms of their growth response. There are many examples in which effects of auxin and of ethylene run parallel, but when ethylene is applied during a growth response to auxin their actions may not be entirely parallel, because ethylene may alter the way the tissues grow in the presence of auxin. For instance, it may foster lateral growth in girth, instead of longitudinal growth. Sometimes applied ethylene seems to give different responses from those expected of auxin, and these have been attributed to the ability of ethylene to reduce the basipetal transport of auxin so that it could induce effects that might otherwise be due to a lack of auxin. There is mounting evidence that ethylene may also inhibit cell divisions in some shoot apices.

Ethylene may originate biosynthetically from methionine (Mapson, 1969; Pratt and Goeschl, 1968; Spencer, 1969; Yang, 1968). Figure 9–5 shows the formation from methionine in such a way that the carbons are numbered to emphasize that the C-1 of methionine gives rise to carbon dioxide, C-2 to formic acid, C-3 and C-4 to ethylene, C-5 to methyldisulfide and methanethiols, and the ammonia group to ammonia. The scheme shown involves a free radical mechanism (not illustrated in detail), which is mediated by light and flavin mononucleotide (FMN), consisting of electron transfer from the sulfur of the methional molecule to the photoactivated FMN followed by a nucleophilic attack by OH^-. Similar mechanisms invoke sulfite-activated peroxidases in which those agents which are produced during univalent reaction of O_2, such as O_2^-, $HO{\cdot}HO_2^-$, and $OH{\cdot}$, substitute for the photoactivated FMN (see Yang, 1968, for details).

Although ethylene, and even acetylene, may induce such profound effects in their own right, the suggestion that all auxin effects should really be attributable to ethylene is too extreme. A study, carried out with Dr. Burg, of ethylene production in relation to the growth induction in carrot explants has produced some interesting results. When the tissue responds to unbalanced applications of growth factors (i.e., to IAA, zeatin, inositol, and the glycosidic active fraction of *Aesculus* as separate components of Systems I and II), substantial bursts of ethylene emerge. Nevertheless,

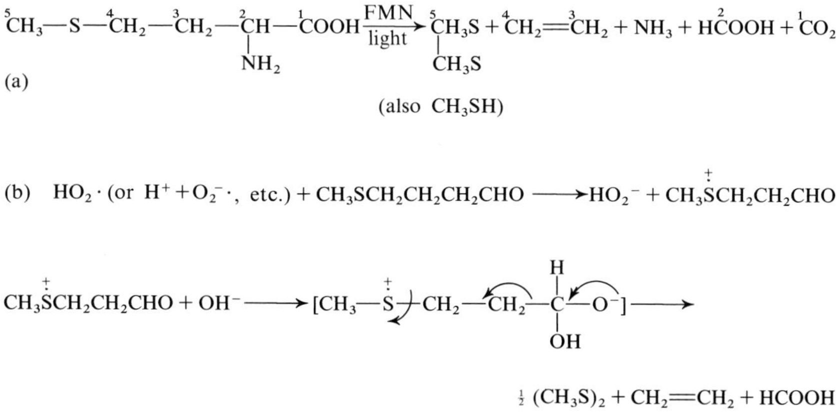

(a)

$$\overset{5}{C}H_3{-}S{-}\overset{4}{C}H_2{-}\overset{3}{C}H_2{-}\underset{NH_2}{\underset{|}{\overset{2}{C}H}}{-}\overset{1}{C}OOH \xrightarrow[\text{light}]{\text{FMN}} \underset{CH_3S}{\underset{|}{\overset{5}{C}H_3S}} + \overset{4}{C}H_2{=}\overset{3}{C}H_2 + NH_3 + H\overset{2}{C}OOH + \overset{1}{C}O_2$$

(also CH_3SH)

$$\text{(b)}\quad HO_2\cdot\ (\text{or } H^+ + O_2^-\cdot,\ \text{etc.}) + CH_3SCH_2CH_2CH_2CHO \longrightarrow HO_2^- + CH_3\overset{+}{\dot{S}}CH_2CH_2CHO$$

$$CH_3\overset{+}{\dot{S}}CH_2CH_2CHO + OH^- \longrightarrow [CH_3{-}\overset{+}{\dot{S}}{-}CH_2{-}CH_2{-}\underset{OH}{\underset{|}{\overset{H}{\overset{|}{C}}}}{-}O^-] \longrightarrow$$

$$\tfrac{1}{2}\,(CH_3S)_2 + CH_2{=}CH_2 + HCOOH$$

Fig. 9-5. The formation of ethylene from methionine (from Yang, 1968). (a) The schematic pathway showing the fate of the carbons; (b) the chain of reactions showing the formation of ethylene from methional (see text).

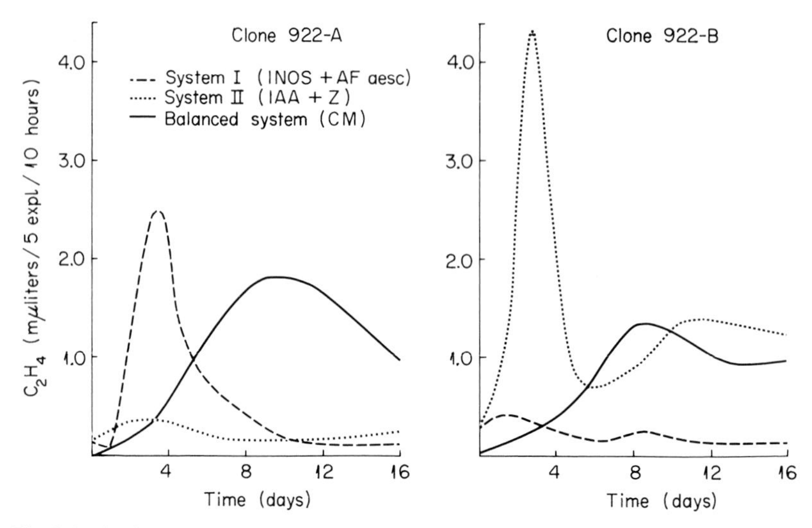

Fig. 9-6. The time course of ethylene output of cultured carrot explants in relation to growth induction (from work of Burg and Steward).

carrot tissue can grow very rapidly under *balanced* conditions of growth factors, such as those in coconut milk, with only very low levels of ethylene production (see Fig. 9–6).

Therefore, it is apparent that the possible responses to ethylene, like those of other prominent growth regulators, are too varied for all of them to fit into one simple direct explanation. (Ethylene-forming substances, for instance, are being used in the rubber plantations to stimulate the exudation of latex.) However, ethylene is conspicuous among plant growth regulators because of the simplicity of its molecule and its ability as a gas to impart its effects widely, not merely from organ to organ but also from plant to plant. But the mode and site of action of ethylene are as obscure as those of any other "hormonelike" substance, although, parenthetically, it is hardly likely to act by forming an integral part of RNA's, or in complexes with nucleic acids (Shimokawa and Kasai, 1968), nor would one expect gene control centers in cells to be directly responsive to a substance so frequently present in traces in the atmosphere.

Concepts Involving Proteins and Nucleic Acids

By analogy with ecdysone, it would seem natural to expect that plant growth regulators should stimulate the activity of certain genes to release specific RNA's into the cytoplasm, such RNA's being able to act as messengers to stimulate the synthesis of specific proteins. So far, the evidence, despite much work (Ingle *et al.*, 1965; Jachymczyk and Cherry, 1968; Key, 1969; Key and Ingle, 1968; Loening, 1968; O'Brien *et al.*, 1968) is not convincing, especially when the systems so tested are not aseptic (Lonberg-Holm, 1967). There are more important objections however.

It has been pointed out that there are very few rapid growth responses in plants besides that of the induction of cell elongation by auxins (see Rayle *et al.*, 1970a, b). Despite the fact that this induction has been intensively studied in the past (see Thimann, 1969), only recently have techniques become available that enable one to continuously monitor the elongation. When auxin is added to a coleoptile system, there is either no effect or only a slight inhibition of growth for a period of from 10 to 15 minutes, followed by a sharper rise in growth rate that reaches a maximum after an additional 15 to 20 minutes (Rayle *et al.*, 1970a, and references cited therein). Dela Fuente and Leopold (1970) have shown that when auxin is withdrawn, there is a continuous effect lasting about 15 minutes, which then declines to 50% after an additional 15 to 40 minutes. The decay has been interpreted as a function of the transport of auxin out of the sections, although for single cells 30 minutes have been estimated for 50% "decay" from the time of auxin withdrawal. These data suggest a loose association–dissociation of auxin within the sensitive sites of the cells.

The lag period of 10–15 minutes necessary for the onset of auxin-induced "growth" can apparently be shortened to 1 minute or less by using the methyl ester of IAA instead of IAA (Rayle *et al.*, 1970a). It is inconsistent to attribute such a short lag period to a result of an auxin effect on gene activity. This is especially so when the induction of β-galactosidase in *E. coli* requires a minimum of 3 minutes (Kepes, 1967).

Evans and Ray (1969) have also followed the kinetics of response to auxin by using a very sensitive, optical method and have shown that the mode of action is incompatible with a genetic target. After treatment with IAA, corn and oat coleoptile segments continue to grow at a slow rate for about 10 minutes. Suddenly, a substantial increase in the rate of elongation occurs. Within an additional 3 minutes, very rapid rates can be observed (about six times that previously noted). The latent period of response of about 10 minutes cannot be extended by pretreatments with actinomycin D, puromycin, or cycloheximide, even though these substances partially inhibit the elongation response (see Noodén and Thimann, 1966). Evans and Ray have presented detailed arguments to indicate that "auxin probably does not act on the elongation of these tissues by promoting the synthesis of informational RNA or of enzymatic protein." They have not excluded, however, "the possibility that auxin acts at the translational level to induce synthesis of a structural protein, such as cell wall protein or membrane protein" (see Ray, 1969, and references cited therein for a detailed summary of this work).

Moreover, Rayle and Cleland (1970) have again emphasized that treatment with substances other than auxins or exposure to CO_2 or low pH can also cause elongation. Low pH (optimum around 3) induces elongation with a lag period of about 1 minute. The low pH must be maintained for continuous response that lasts only from about 1 to 2 hours. The response to CO_2-saturated water is similar to the effect of pH, but the maximum growth rate is far greater than that obtained with auxin.

A curious feature of the low pH response is that it works on cells of *Avena* coleoptiles that have been frozen and thawed (see Rayle *et al.,* 1970b). This suggested that the pH response and the response to auxin involves a common "cell wall loosening process" (Rayle *et al.,* 1970b). This leads to serious implications about the concepts involving cell walls in growth regulation, because it means that cell wall synthesis is not directly involved in the control of cell elongation and, hence, the action of auxin on cell wall synthesis is not applicable. Nevertheless, (Rayle *et al.,* 1970b) tentatively suggest that auxin might induce elongation by facilitating the release into the cell wall of hydrogen ions, or of some enzyme capable of breaking acid labile cross-links of the cell wall.

IAA–RNA complexes [more specifically IAA thought to be bound to a light (4 S) transfer RNA] have also been invoked (Galston, 1967b), although without much chemical justification; the idea has now been abandoned by Galston and Davies (see Key, 1969, and references cited therein) after they obtained separation of the labeled IAA fraction from the RNA.

IAA oxidases (peroxidases) have been considered as regulators of IAA levels in older tissues, with the IAA actually "triggering" the production of the enzyme to which it owes its destruction (Galston, 1967b; Galston *et al.*, 1968; Pilet and Gaspar, 1968; Ray, 1958). This enzyme, indoleacetic acid oxidase, converts indole-3-acetic acid to 3-methyleneoxindole (see Fig. 9-7), and despite earlier criticisms that some auxins could survive in plants rich in the oxidase Galston still believes that the IAA oxidase may function *in situ* in a regulatory way (Galston, 1967b; Galston and Davies, 1969, 1970).

Devices such as those made familiar under the principle of metabolic feedback offer possibilities of inducing drastic change in cell metabolism. For example, feedback inhibition or activation of enzyme activity by intermediates and end products of a pathway involve rapid interactions between small molecules and macromolecules. The time required for such a system to return to the equilibrium state once it has been disturbed is very short (of the order of a few milliseconds, or for complex systems, a few minutes at most). The binding of an "effector" molecule to a site other than the active site of the enzyme can greatly change the relationship between the speed of the reaction and the substrate concentration. The response of the enzyme to such binding has been termed an allosteric affect (Monod, 1966; Monod *et al.*, 1965). The effector molecule may be an end

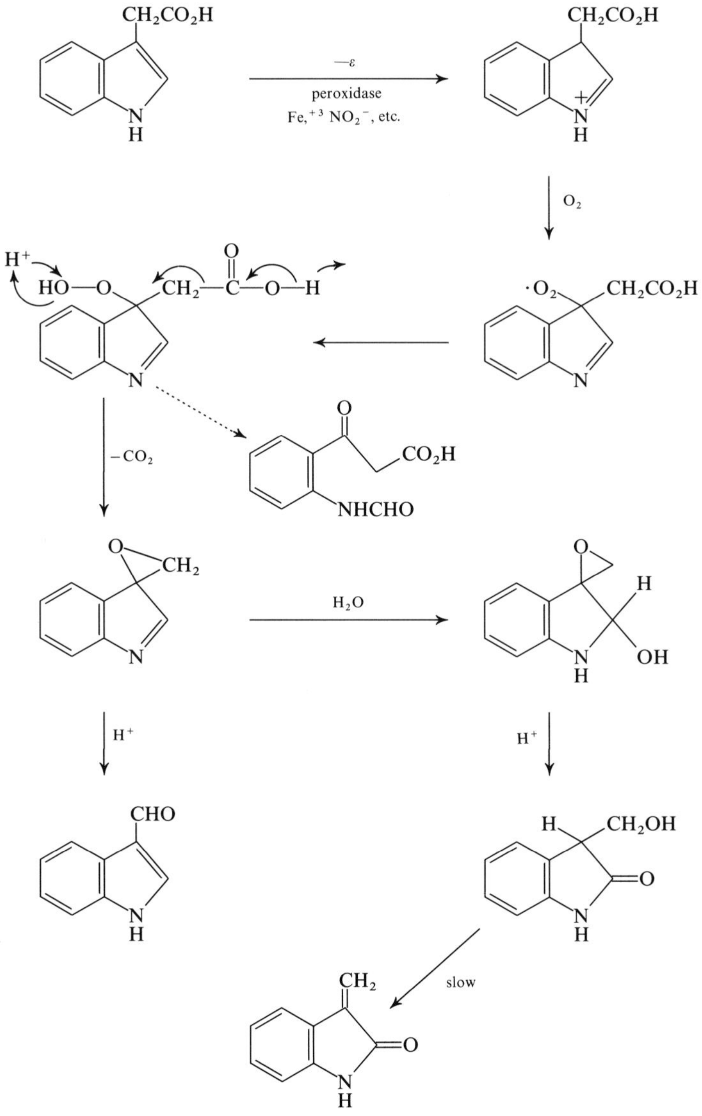

Fig. 9-7. Oxidation of IAA by IAA oxidase to 3-methyleneoxindole (Hinman and Lang, 1965).

product, or an intermediate in the pathway, or some other small molecule; it may activate rather than inhibit, and there may be more than one effector for a given allosteric protein.

Some feel that indole-3-acetic acid acts as an allosteric effector (Dahlhelm, 1969). Heavy water (D_2O) inhibits auxin-induced cell elongation in grass hypocotyls and also prevents the formation of indoleacetic–aspartic acid. Since the synthesis of IAA–aspartic acid is enzyme-mediated, this is taken as evidence that D_2O has an effect on the production or activity of the enzyme. Since the addition of IAA–aspartic acid had no effect on growth of the grass hypocotyls in the presence of heavy water, and the protein patterns upon electrophoresis were the same in H_2O and D_2O, it has been suggested that the D_2O has a stabilizing effect on allosteric proteins.

There is evidence that IAA is a positive allosteric effector of citrate oxalacetate lyase (CoA-acetylating), the enzyme that catalyzes citrate + CoA → acetyl CoA + H_2O + oxalacetate. The view here is that IAA produces a "signal" by means of an electron which it may lose; the receptor of this electron could be the sulfhydryl of proteins. Since the sulfhydryl is important in determining protein conformation, a change in conformation might be effected by the acceptance of an electron by SH of the citrate oxalacetate lyase, and thereby IAA becomes involved, according to Sarkissian (1968), in "RNA synthesis, protein synthesis, oxidative phosphorylation, protoplasmic streaming, loosening of the cell wall," etc. But, even so, the primary observations of Sarkissian are called in question by Zenk and Nissl (1968) who measured the effect of IAA, 2,4-D, and NAA upon the action of highly purified citrate synthase without any positive effect whatsoever. Therefore, if indeed the allosteric effects mediated by these auxins are real, they involve other substances that were present in the impure preparations. Along the same lines, abscisic acid has been implicated by van Overbeek *et al.* (1968) as an inhibitor in the scheme of allosteric effects. Obviously the involvement of "allosteric effects" in the action of growth regulators is both highly specific and equally speculative.

Nuclei isolated from dwarf pea shoots show a higher rate of DNA-dependent RNA synthesis in the presence of gibberellic acid. The RNA synthesized by the treated nuclei showed a different "nearest-neighbor" frequency. It differed in molecular weight from the RNA synthesized by control nuclei, and suggested that gibberellic acid "can thus modify the RNA synthesized by isolated nuclei" (Johri and Varner, 1968). Gibberellic acid has also been invoked as an agent that primarily acts "upon membranes and transport mechanisms" (Lüttze *et al.*, 1968). It has also been suggested that gibberellic acid exerts its effects by working on the mitotic apparatus (Fragata, 1970). This view derives from the observation that gibberellic acid is able to partially reverse colchicine-induced inhibition of coleoptile elongation, and is based on the fact that colchicine acts by suppressing spindle formation during the development of the mitotic apparatus by attaching to the subunit proteins of the microtubules (Shelanski and Taylor, 1967; Deysson, 1968).

Some of the growth retardants such as AMO 1618, Phosphon S, and Phosphon are potent inhibitors of the enzyme kaurene synthase. This enzyme is responsible for the conversion of *trans*-geranylgeranylpyrophosphate to *ent*-kaurene. The observed action of these retardants (see Fig. 7-5) is thus conceived to be due to the inhibition of gibberellin synthesis (Paleg *et al.*, 1965; West and Upper, 1969). None of this detail, however, leads to a satisfying molecular mechanism at a precise site of action.

The idea, previously discussed (see page 132 *et seq.*) that a given class of growth substances, e.g., the adenyl compounds, could act by entering into combination with tRNA's or somehow influence the "pool" of tRNA species (Hall, 1970; Skoog and Leonard, 1968; Skoog and Armstrong, 1970) and so affect protein synthesis, though interesting, has not as yet gained general credence.

The protein synthesis inhibitors [e.g., cycloheximide, actinomycin D, etc. (Reich and Goldberg, 1964)] offer understanding at a specific molecular level, although these substances are best known in microbial systems. These and other substances are now known to intervene at distinct stages, e.g., cycloheximide mainly suppresses the lengthening of the peptide chains, while actinomycin D acts as a RNA polymerase inhibitor (see Waring, 1968, and references cited therein). But bacterial antibiotics have had less use in higher plants because of problems inherent in their entry into the cells and their penetration to active sites (see Brian, 1957; Ellis and MacDonald, 1970; Key, 1969, and references cited therein).

Some nonprotein amino acids, which can act as analogs for protein amino acids, may interfere with protein synthesis by competitive inhibition with the protein amino acid concerned (Fowden *et al.*, 1968; Luckwill, 1968).

In a similar way purine and pyrimidine analogs, or substances that can intervene in purine and pyrimidine biosynthetic pathways, may have effects of RNA and DNA synthesis because they affect the formation of the necessary bases (Bücher and Sies, 1969; Fishbein *et al.*, 1970). Such substances would be expected to influence cell division (Nickell, 1955). Thus, 5-bromodeoxyuridine (BUdR) and 5-fluorodeoxyuridine (FUdR) result in a deficiency of thymidylic acid and, hence, block DNA synthesis. Similarly, aminopterin (4-aminopteroylglutamic acid) inhibits folic acid reductase, which catalyzes tetrahydrofolic acid synthesis. Since the tetrahydrofolic acid is a coenzyme in one-carbon transfer reactions (involved in purine syntheses) and in the formation of thymidylic acid from deoxyuridylic acid, its inhibition again inhibits RNA and DNA synthesis, and, hence, cell division (see Deysson, 1968; Fishbein *et al.,* 1970; Kihlman, 1966, and references cited therein; Krieg, 1963; Singer and Fraenkel-Conrat, 1969). Along these lines, 5-aminouracil which inhibits DNA synthesis, has been used to block cell divisions in root tip meristems, and, by its subsequent removal, to foster synchronized cell divisions (Clowes and Juniper, 1968).

Acridines and phenanthridines are examples of substances which can combine with nucleic acids and bring about important physical changes in their molecular structure (Waring, 1968). In the case of DNA, proflavine molecules are thought to be intercalated between adjacent base-pairs of the double helix, thus making the molecule longer and more slender (Lerman, 1961; Waring, 1968). It is to be expected, therefore, that these substances have important consequences for mutagenicity and carcinogenicity (Krieg, 1963; Singer and Fraenkel-Conrat, 1969).

Concepts Involving Light and Phytochrome

In any hormone theory of perception and transmission of a stimulus some molecule must play a key role. Where the stimulus involves light (see Fig. 9-8), phytochrome is the favored perceptor and when activated, as by the absorbance of red light P_R is transformed to P_{FR} (see Black, 1969; Furuya, 1968; Hendricks and Siegelman, 1967; Hillman, 1967; Siegelman and Hendricks, 1964). [No attempt will be made here to go into details of this well-known system. The morphogenetic changes induced by red light have been extensively reviewed by Galston and Davies (1970).] A great deal of sophisticated chemistry has gone into the study of the phytochrome molecule and it now seems that at least four species of phytochrome chromophore, having absorption maxima at 580, 600, 670, and 730 mμ, are present in rye phytochrome (Correll *et al.,* 1968). The structure proposed for a phytochrome chromophore with its protein attached is shown (see Fig. 9–9; Rüdiger, 1969; Siegelman *et al.,* 1968).

Although most of the known phytochrome-mediated systems (see Table 3–1, page 34) show changes requiring days or at least hours, the general view is that the primary or initial effect of irradiation on phytochrome is probably a rapid one followed by secondary reactions. Permeability changes in response to red–far-red have been studied (see Galston and Davies, 1970; and references cited therein) using the sleep movements of *Albizzia* and *Mimosa* as the indicator. The pinnules of both these genera fold together in darkness; in light they are open. On being treated with red light (thus converting P_R to P_{FR}), the permeability of the membrane is so altered

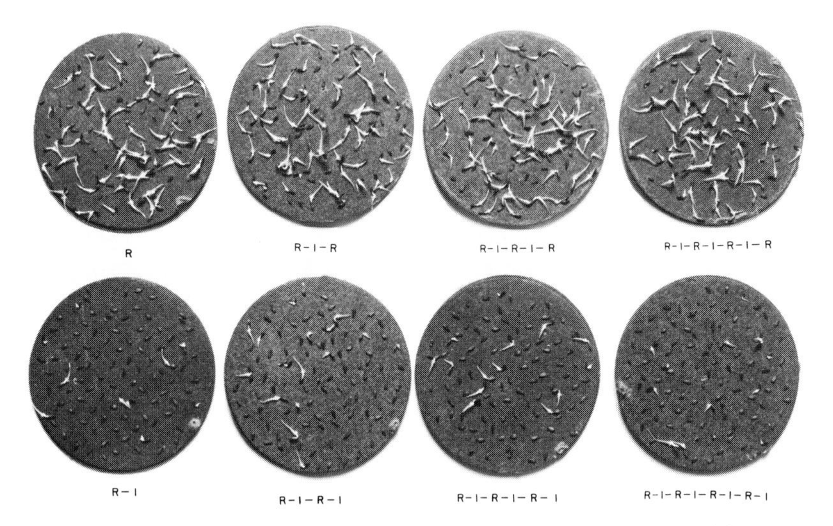

Fig. 9-8. Effects of red (R) and far-red (I) light mediated by phytochrome and demonstrated by germination of lettuce seed (*Lactuca sativa* var. Grand Rapids). Upper row, seeds germinated since their final treatment in the alternating series was with red (R) light; lower row, few seeds germinated since their final treatment in the alternating series was with far-red (I) light. (Photograph courtesy of the U. S. Department of Agriculture.)

that rapid closure of the pinnules occurs in darkness. If the plant is first treated with far-red light prior to being placed in darkness, the pinnules remain open. Since these sleep movements are known to be related to the turgor of the pulvinules (the leaves are open when they are turgid and closed when they are flaccid), it seems reasonable to assume that the phytochrome system is located in functional membranes.

Based on the orientation of the flat, platelike chloroplasts in *Mougeotia* to precisely directed incident light, Haupt (1968) conceives that the chromophore is in the membrane surface of the alga, and he attributes the effect of the light to the orientation of the photoreceptor molecules in relation to the cell surface and to the incident light.

Köhler (1969) has associated phytochrome with ion transport. He has shown that red light inhibits the uptake of potassium into internodes of pea seedlings and promotes uptake into the plumules.

Tanada (1968) has made the observation that root tips of some species of plants adhere to glass surfaces when they are irradiated with red light (660 mμ) and detach from the glass within 30 seconds when they are irradiated with far-red light (730 mμ). Thus, this rapid response is characteristic for phytochrome-mediated reactions. Yunghans and Jaffe (1970) have studied adhesion and release in detail and have shown that the glass surface may be negatively charged with thiocyanate (SCN^-), nitrate (NO_3^-),

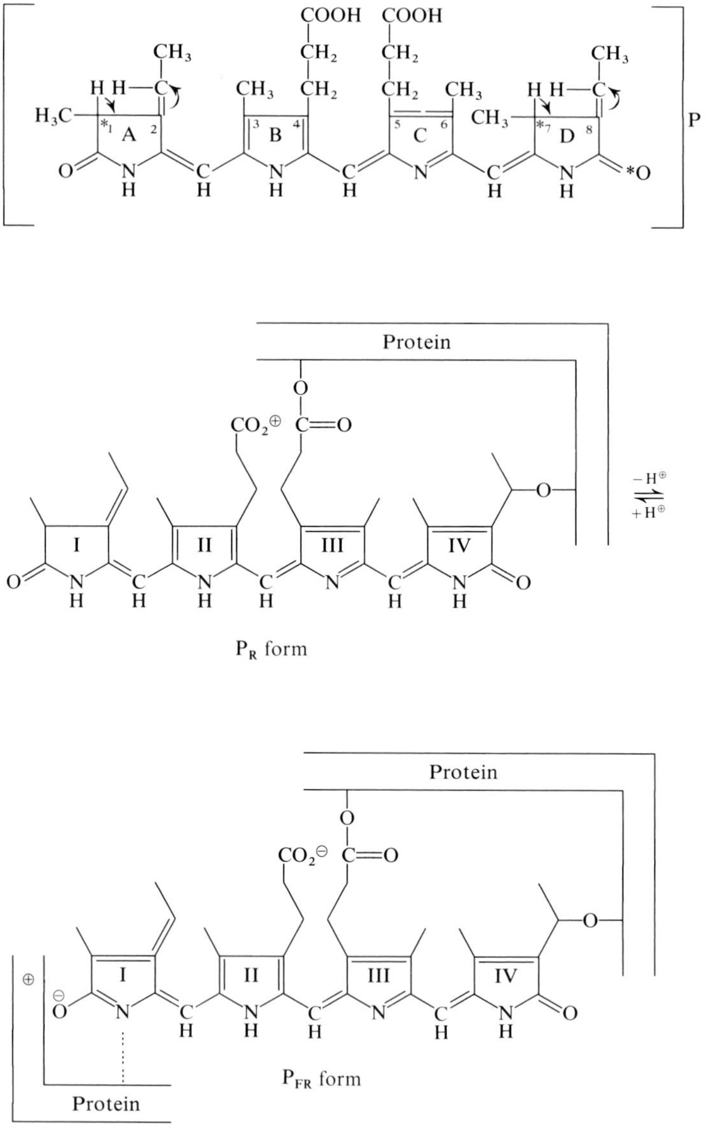

P_R form

P_{FR} form

Fig. 9-9. Proposed structures for phytochrome chromophore. *Top*, the structure after Siegelman *et al.* (1968). Here the attachment to the protein is unspecified; the emphasis being upon the intramolecular rearrangement that could increase the number of conjugated double bonds from 7 to 10 when P_R is converted to P_{FR}. *Center* and *Bottom*, the structure after Rüdiger (1969), with pigment bound to protein and showing the $P_R - P_{FR}$ interconversion by light. Note that the identical ring systems are differently identified in these formulas.

sulfate (SO_4^{-2}), chloride (Cl^-), phosphate (PO_4^{-3}), citrate $C_6H_5O_7^{-3}$), oxalate ($C_2O_4^{-2}$), or glutamine ($C_5H_8NO_4^-$). Benzoate ($C_7H_5O_2^-$) and acetate (CH_3COO^-) were relatively ineffective for red light adhesion; however when citrate and oxalate were used release was inhibited. These investigators felt that this was due to chelation of Ca^{2+}, because release began simultaneously with the addition of excess Ca^{2+} to the bathing solution under far-red light. They conclude that "chelation may remove Ca^{2+} essential for the functioning of a selective ion-transport system, since calcium is important in the selective uptake of other essential ions by roots."

It thus seems clear that there is evidence that phytochrome may indeed be localized within the membrane. Nevertheless, Rubinstein *et al.* (1969) have searched for a membrane associated with phytochrome, but have found that the bulk of the phytochrome in etiolated oat seedlings is distinct from that which was associated with sedimented pellets. The amounts found associated with membranes could, however, be of physiological significance (Rubinstein *et al.*, 1969), and could explain the fact that only a portion of the total phytochrome in a plant mediates the red–far-red effects.

This chromoprotein has also been endowed with enzymatic properties (Hendricks, 1959).

Whereas Galston (1968) attempted to localize phytochrome in cells of *Avena* coleoptiles and pea internodes (stressing the nucleus as a principal site), other investigators who have described its distribution in the embryos of Cucurbitaceae (Boisard and Malcoste, 1970) stress its greater prominence in the cotyledons than in the vegetative axis, and point to further studies to locate it at more specific sites of action during germination. The view that it may act by regulating gene repression and derepression and thus ultimately act through mRNA-mediated mechanisms to bring about positive photoresponses has also been espoused (Mohr, 1966, 1969). Suggestions have also been made that red light stimulates the synthesis of several enzymes (Graham *et al.*, 1968; Tezuka and Yamamoto, 1969; Weidner *et al.*, 1968), or even gibberellins (Reid *et al.*, 1968; Reid and Clements, 1968), and that far red stimulates enzyme activity, e.g., that of phenylalanine ammonia lyase—the enzyme which deaminates phenylalanine to yield *p*-coumaric acid (Smith and Attridge, 1970; Weidner *et al.*, 1968, 1969).

Riboflavin-like substances have also been credited with light-perceptive abilities (Davis, 1968; Galston, 1967b; Galston and Davies, 1970; Thimann, 1967a). Thimann drew attention to certain crystalline bodies which were candidates for photoreceptor sites in oat coleoptiles and in sporangiophores of *Phycomyces* (see Thimann, 1967b, and references cited therein). However, these bodies seem now to be more ubiquitous than this role would demand (Frederick *et al.*, 1968).

Flowering responses of plants are strikingly linked to light periodicities, both short- and long-day plants being known (see Lang, 1965). It is in this area that the presumed role of phytochrome could be so important. But it is now even suggested, with some evidence, that the phytochrome system may also function as a perceptor in thermally regulated events (Anderson *et al.*, 1969).

Claims have been made again that florally induced *Xanthium* buds possess characteristic RNA profiles (Cherry, 1970; Cherry and van Huystee, 1965), and the importance of the RNA in flowering has been emphasized by the work of Heslop-Harrison who utilized the pyrimidine, 2-thiouracil, which interfered with RNA metabolism and this, in turn, prevented flowering [see Heslop-Harrison, 1960; and the volume edited by Bernier (1970) on "Cellular and Molecular Aspects of Floral Induction" for detailed discussion on this subject]. Even the sex of flowers, of a monoecious plant (*Cucurbita*) and a dioecious one (*Cannabis*), has been brought under chemical control by application of certain growth factors, e.g., NAA application on the former yields a high proportion of female flowers, and, in the latter, plants of the genetically male sex responded by forming female flowers! Gibberellic acid may also accentuate the trend toward maleness (Heslop-Harrison, 1964).

Again, it should be noted, even in the above selected examples, that the descriptions of the responses due to the growth regulators far outweigh the strict interpretation of their mechanisms at the molecular level.

CHAPTER 10 **Prospects and Problems**

It should now be very clear that great and often unexploited potentialities are inherent in the controlled chemical regulation of the growth and behavior of plants. The difficulties in seeking ultimate explanations of these actions have been stressed. One can describe the responses that the chemical substances elicit, although one may not be able to say precisely why they do what they do. This is a familiar dilemma in the understanding of problems which relate specifically to the highly organized systems which are alive. We cannot yet write out the prescription for all the subtle features, chemical and physical, which distinguish the living from the nonliving state. We cannot even say "why," as distinct from "how," such molecules as DNA replicate in their milieu in plant cells. The dilemma is that we are trying to understand, as distinct from describe, how a great array of substances, natural and synthetic, intervene to modify a complex system which is, in this context, not defined.

The inherent difficulty in problems of chemical plant growth regulation lies in the very small amounts that are needed to produce profound effects. This should not be surprising. Cells are proverbially small (carrot cells are of the order of 0.10–0.15 μg), but although small in terms of size, they are complex in terms of numbers of molecules (3×10^{15} water molecules; 1×10^{9} small protein molecules). Although needs for major nutrients for whole plants may seem impressive, the particular role of such special ones as the trace elements may require amounts so small as to raise the question: How many atoms of a given element, e.g., Mo, can suffice for the needs of one cell? The entire genetic plan is transmitted from cell to cell by a very small amount of nuclear DNA (order of 15×10^{-6} μg as for diploid cells of corn). To carry the argument somewhat further, the individual bases, whose location on the DNA is critical, are present in a given cell in even smaller amounts. Hence, the extraneous molecules whose role is regulatory rather than nutritional may intervene to provoke effects by the use *per cell* of incredibly small amounts if these are expressed in usual chemical terms. The molar concentration of indole-3-acetic acid (IAA) ranges from 10^{-9} to 10^{-3} when applied to shoots and from 10^{-11} to 10^{-7} when applied to roots (Thimann, 1948, 1952). These already minute amounts should be related to the many cells they affect and allowance should also be made for the inefficiency of external application, since further dilution must occur before reaching the actual site of action. One is dealing here with the use of molecules by plants (see Chapter 7) which should properly be interpreted not on their gross application to areas of ground or to plant surfaces or even to cells in culture media but, for precision, should be expressed as the number of molecules needed at specified active sites to bring about a given amount of biological response. It is obvious that the problems cannot yet be analyzed fully in such precise terms. Some attempts have been made, however, to estimate the number of auxin molecules active per growing cell (see Weevers, 1949, p. 186). For example, in the *Avena* coleoptile the number has variously been placed at 10^4 or 3.6×10^4, whereas in isolated maize roots 4×10^5 molecules were estimated to be active per cell. These numbers are small when compared with the number of protein molecules in a cell. If 68,000 is taken as the average molecular weight of a protein, then one molecule of auxin is active for, roughly, between 10,000 and 1,000,000 protein molecules (Weevers, 1949). Ideally one should contrast the nutritive molecules which are needed in amounts that require the statistical treatments of solution chemistry with the regulatory molecules of which only a few per cell are needed to activate precisely defined sites (Matsubara *et al.*, 1968).

Therefore, to apply chemical growth regulators externally and in small amounts to plant cells is still like shooting projectiles into a molecular universe without knowing what targets they hit or knowing where or how they act (see Worley, 1968). Moreover, the device, currently practiced of testing a chemical mechanism that underlies a biological response by demonstrating its effectiveness *in vitro* is not feasible here, or even appropriate. (The often used reference to growth of plant parts *in vitro* is, in fact, entirely inappropriate, for the growth occurs in the cells, *not* in the containers!)

It is not surprising then that plant cells, organs, and plants emerge as systems capable of multiple responses to exogenous chemical compounds, since innumerable sites react in innumerable ways. For this reason, the simplistic or unitary approaches so profitably practiced in modern genetics, or in enzymology, are here of little avail. The unit, enzyme-mediated steps, such as those shown in the increasingly elaborate metabolic charts that adorn laboratory walls, may tell us what is chemically feasible in organisms but they do not remotely tell us how all this biochemical machinery is coordinated and regulated in the work of a cell. The biochemical machinery is subjected in cells to restraints and controls imposed when the reaction systems become part of its highly compartmented organization.

The Complexity of the System: Its Multivariate Responses

The challenge of the chemical plant growth regulators is that they can intervene in this complex system to determine how it works in a balanced or imbalanced way. And, because living plant cells are essentially totipotent, because they are essentially adapted to live in free contact with the outside world and are not as isolated as are cells of higher animals in a very specialized, protected, internal environment, they make use of, and respond to, the stimuli that may be mediated by a vast array of physiologically active molecules. But the attempt to cast these molecules into rigidly defined classes in terms of what they do in highly specialized assay systems, or to interpret their role too dogmatically on the classical hormone model as it applies to higher animals, loses sight of essential characteristics of plant cells. Moreover such points of view do not rise to the distinctive challenge of the problem in plants, nor do they encourage the search for responses to multiple factors acting concomitantly on systems that are fully able to grow.

Simplicity is always attractive *if* it can be achieved. But the history is clear. From *the* growth substance, *the* auxin, the known active molecules proliferated; from *one* class of growth regulators *many* have now emerged and, recently, the counterparts of the single factor have multiplied with incredible speed, as the record clearly shows. The subject of the chemical regulation of plant growth, that is, the study of all the exogenous chemical triggers which put the otherwise idling metabolic machinery to work in directed ways, should now enter a new phase.

In this phase, the overriding impulse should be to investigate not the system that is so restricted that its responses are predictable, but rather the most complete and totipotent ones that can respond fully to the interplay of externally applied factors; and to investigate, not merely the single regulatory factor in isolation when its action may be limited, but to do so in relation to all others that can by interactions, synergistically and sequentially, bring out the full range of biological responses which the chemical growth regulators may elicit. The observed responses should not be restricted to the obvious ones, e.g., overall criteria of growth by size or weight, for a full understanding may only come from the use of criteria applied at all levels from those of gross morphology to metabolism and fine structure of cells (Steward *et al.*, 1968a).

Geneticists face similar dilemmas as they become familiar with the ways an organism may inherit and transmit the innumerable units that establish its detailed characteristics without their being able to prescribe a master unit that decrees that the organism, or its organs, shall even exist as such. One has to recognize that, with increasing complexity of biological systems, culminating in those that can grow and develop, they possess characteristics that are over and above the sum of their parts. Conversely, the component parts have a greater range of effectiveness, and even different and new regulatory roles when they function as part of the whole. Those who seek to interpret the role of growth regulators in plants deal not only with substances that exert some specific effect at a selected site, but perhaps, even more, must acknowledge their resultant ramifications through a balanced network, a matrix of interlocking systems constituting the living system that can grow.

At this point we need means to visualize the multiple consequences of even a single variable in the growth, nutrition, and metabolism of plants, and we need to study this on *whole* systems that can grow, and not only upon their isolated parts. This is essential to understanding, or measuring, the many interactions which flow from a given treatment.

We have long been familiar with the morphological consequences of length of day (photoperiodicity); this may be perceived by phytochrome and, as a first step, mediated by its activated form (P_R) (Hendricks and Siegelman, 1967; Hillman, 1967). Earlier concepts that the composition, metabolism, and nutrition of photoperiodically stimulated plants were not changed are not valid, as work on the mint plant showed as early as 1950 (see Steward, 1968, and references cited therein). But how can one visualize the multiplicity of effects that flow from a single variable like length of day, probably attributable ultimately to a single causative agent? The polygonal diagram of Fig. 10-1 shows how the consequences of the impact of this variable may be seen in terms of a network of effects; these range from the content of single substances in leaves to the form of the whole plant and to the way the plant uses its nutrients.

The growth responses of carrot explants to growth stimuli and the idiosyncracies of different clones have already been explored (see page 64). These show how the endogenous capacity of the carrot cells to grow responds to the partial components of a total growth factor system (IAA or inositol) and later to a more complete system (as in coconut milk). The specific responses of different carrot clones affect not only the growth in terms of substance (fresh weight) but also the way the cells discriminate

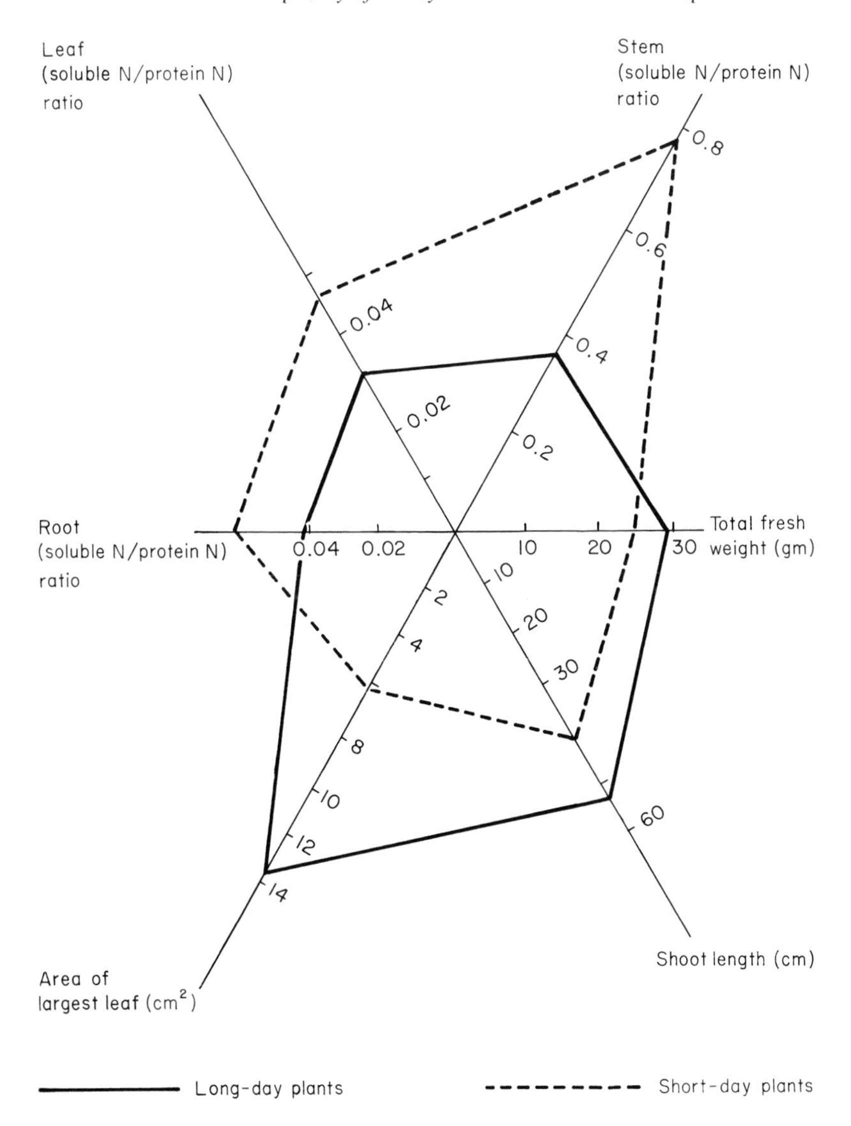

Fig. 10-1. Polygonal diagram to show the interactions of environmental factors with the growth and composition of mint plants grown in nutrient solution (see Steward, 1968).

between water and salts (see Fig. 10-2), even as the nature of the growth factors employed also affects the composition of the resultant cells (Steward and Mott, 1970). Thus, it is not a retrograde step to pay attention to clones with distinctive properties, as some may have indicated, If the objective is to understand the potentialities of a given genotype and the full range

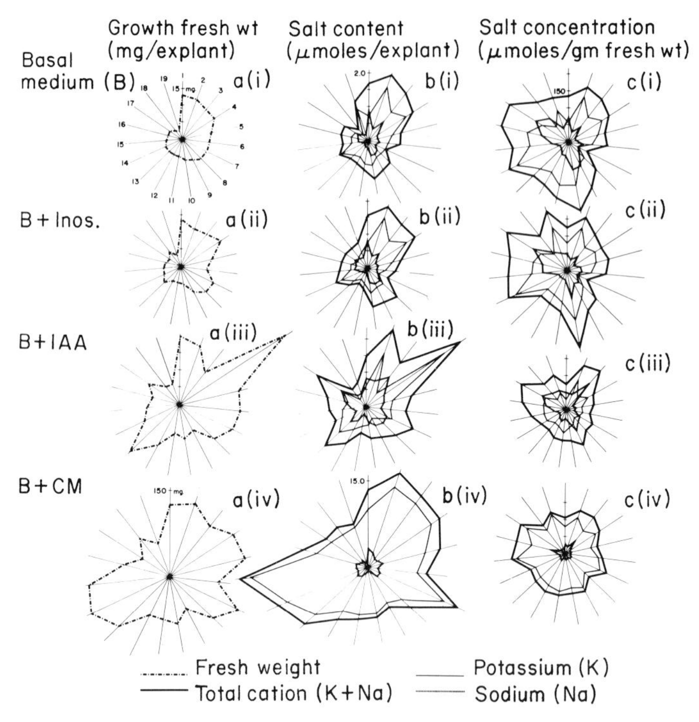

Fig. 10-2. The responses of 19 different carrot clones to different stimuli as indicated by their growth in fresh weight and their salt content per explant and per gram fresh weight. Series a(i) shows the growth response in the basal medium in descending order. Series a(ii–iv) shows the responses of the same clones to growth-promoting substances (IAA, inositol, and coconut milk). The scales for a(iv) were reduced to accommodate the large growth in fresh weight that occurred. Series b(i–iv) and c(i–iv) show the corresponding effects on total salt content or concentration to reveal the effects of the treatments and idiosyncrasies of the clones (Steward and Mott, 1970).

of its responses to applied conditions, one learns much from the contrasting behavior of different strains, or clones, or even of different seasonal crops. But if this task is too baffling, it may be rendered simpler by so restricting the potentialities inherent in the plant material that it predictably yields but one kind of result, and as effectively ensures that other, equally instructive responses of the plant in question will remain undetected.

The very multiplicity of responses to exogenous chemical growth regulators points to the essential contrast with animal hormones. An animal hormone may be produced at an internal site specialized for the purpose; it may act at a site rendered receptive to the one hormone. Even some of the best documented examples of the action of an animal hormone, e.g., epinephrine, do not seem to require the intervention of *de novo* protein synthesis (see Jaffe, 1969, and references cited therein). Though animal hormones are various in nature and modes of action (Tata, 1968), the effects elicited by them in the organs they affect may be transmitted by increased levels of cyclic AMP (Robison *et al.*, 1968). Recent evidence suggests that binding of cyclic AMP to a protein is required for cyclic AMP activity (Emmer *et al.*, 1970). In other words, the animal hormones as "triggers" are highly specific and the organ affected has a limited range of response; but there may be, even here, a somewhat general use of an intermediary molecule.

The involvement of cyclic adenosine-3′,5′-monophosphate (AMP) in cell communications and the activity of animal hormones have been reviewed by Rasmussen (1970). Many animal hormones stimulate production of cyclic AMP. The idea is that a hormone reaches its site of action from its source and activates the plasma membrane surface by promoting the production by the cell of the enzyme (adenyl cyclase) producing the cyclic AMP in the cell. As part of this activation, calcium ions, essential for the enzyme's activity, encounter a more permeable membrane. Thereafter the cyclic AMP causes proteins in the cell to be phosphorylated. It has been shown that cyclic AMP activates protein kinase which phosphorylates RNA polymerase (see Martelo *et al.*, 1970). The view projected by this work is that RNA polymerase then stimulates RNA synthesis which then directs protein synthesis. The crucial point is, however, whether or not cyclic AMP functions as a trigger in this way in plant cells, and so far, the evidence does not yet exist to connect it with the action of plant growth regulators (see Duffus and Duffus, 1969). Plant growth regulatory systems, moreover, do not have this degree of specificity. The regulators act upon cells at large, and in doing so many set in motion very diverse events. As shown, the responses elicited may even culminate in the totipotent growth of cells which develop into plants.

Adenosine 3′,5′-phosphate
(cyclic 3′,5′-AMP, cyclic adenylic acid)

Even though a given plant growth regulator may demonstrably activate a given gene, even though it may act primarily by stimulating a given enzyme (Feierabend, 1969; Filner *et al.*, 1969), its final consequences would only emerge as its effect disturbs and ramifies through, the balanced network of interlocking relationships which even a single cell presents.

While attempts to detect the first and unit responses of cells, organs, or organisms to exogenous chemicals should proceed, there should also be studies that consider the whole economy of the cells and plants affected. The days of the single factor response in the analysis of growth virtually seem over (see Thimann, 1956) for the ability to grow arises from the interactions between many factors, and a level of organization which permits such interactions constitutes the system that is alive and that can develop. Seen in these terms, the modern investigation of growth-regulating substances is rarely a task for one person; it requires the resources of a team. It also demands an analytical approach beyond the resources of those who conventionally think at one time in terms of one response by one reaction to one factor. Such a standpoint may suffice to interpret chemistry in a test tube, but it can hardly serve for the responses of a cell that grows into a plant. If the parameters of so complex a system can be recognized, biologists may then hopefully expect to draw upon systems analysis or developments in matrix theory in its interpretation.

Growth Regulators and the Tumor Problem

Despite the enormous amount of evidence upon the chemical nature of carcinogens and the responses they elicit, there is no universally applicable and simple theory to tell why cancer cells grow *in situ* (Braun, 1969a,b, 1970). Perhaps there is a lesson to be learned here from plants. Because plant cells may grow in isolation to plants and survive in much simpler media than animal cells, some relationships that are obscured in cultured animal cells emerge more clearly in plants. Instead of seeking some magic single characteristic of cancer cells that will distinguish them in action from their normal counterparts, one should perhaps concede that the differences will be multiple. The cancer cells may embark on their self-perpetuating growth in response to stimuli that produce a highly improbable situation. In terms of the cell cycle this would mean that all the steps could work so that mitosis (M) is uninhibited and that all factors that normally intervene to cause subsequent differentiation, quiescence, or senescence are nullified.

Braun and his co-workers (Wood *et al.*, 1969) have found that plant tumors

Fig. 10-3. Reversibility of the tumorous state. A, Culture of unorganized crown-gall tumor tissue; B, result obtained when a fragment of sterile tumor (see A) was grafted to the cut stem end of a tobacco plant; C, culture of crown-gall tobacco teratoma tissue of single cell origin; D, result obtained when a fragment of sterile teratoma (see C) was similarly grafted. E, F, and G show three stages in the full recovery of a crown-gall teratoma. E, Tumor bud, as shown in D, was grafted to the cut stem of a normal tobacco plant. Note the abnormal resulting growth; F, a tumor shoot of the type shown at E was grafted to the cut stem end of a normal tobacco plant; the resulting growth was more normal than that shown at E. G, Seed from a recovered scion such as that found at F was germinated and gave rise to normal plants. [Photographs supplied by Dr. A. C. Braun, Rockefeller University (Braun, 1961).]

produce a complex substituted purinone derivative (see page 90 and Wood, 1970). However, this "cell division factor" is not the *cause* of tumor formation but the result of the transformation of the normal to the tumorous state. All cytokinin-stimulated cells, about to divide, are thought to produce the same substance. In terms of the cell cycle, this "cell division factor" works at stage (M), whereas it is thought that other substances (adenyl cytokinins, etc.) may prepare the cells for, but are unable to consummate, cytokinesis. It is anticipated that the first member of this special class of "cell division factors" may well be the forerunner of many (Wood, 1970).

One should not forget, however, that some investigators would explain the properties of cancer cells without invoking either the nucleic acids or the cyclical events of the cell cycle, and merely emphasize their altered membrane surfaces (see Hause *et al.,* 1970). Moreover, the tumorous state may be reversible (see Braun and Wood, 1969, and Fig. 10-3).

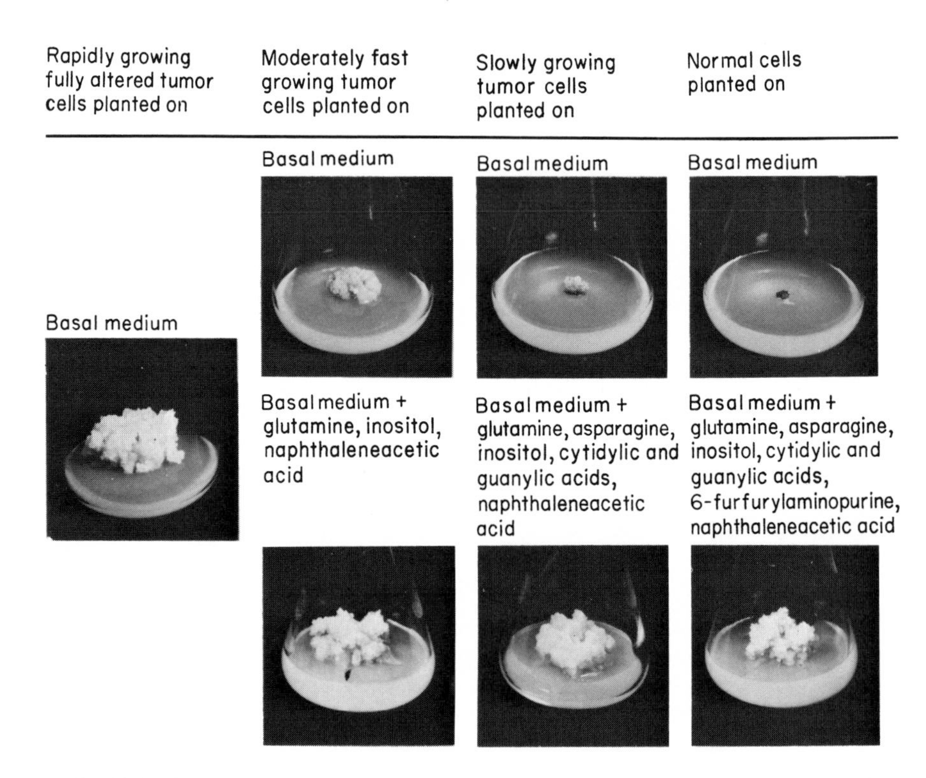

Fig. 10-4. The autonomous growth of fully transformed tumorous cells and the effect of nutrients and growth substances on partially transformed cells. At the extreme left, fully transformed cells are shown growing on a simple medium. Normal cells and cells at various stages of transformation are also shown on the basal medium with different supplements. [Photograph supplied by Dr. A. C. Braun, Rockefeller University (Braun and Stonier, 1958).]

The problem of tumors illustrates the difficulties encountered even when one is concerned with chemical growth stimulation solely at the level of cells that do not form organized cell patterns or develop into plants (see Braun and Stonier, 1958, and Fig. 10-4). How much more, therefore, is this true when growth regulation must also comprehend the complexities of the control over cells as they differentiate into tissues and organs; the control over plants in their environment (prominently affected by length of day and night temperature); the control of symbiotic relationships as it occurs in nitrogen-fixing nodules (Dixon, 1969; Hardy and Knight, 1968), plant galls (Darlington, 1968), mycorrhizae (Gerdemann, 1968), or even lichens (Ahmadjian, 1967; Mosbach, 1969); and as it even extends to the relationships between members of naturally occurring populations via allelopathic substances (Evenari, 1961; Grümmer, 1961; Muller, 1966, 1970; Rice, 1967; Winter, 1961)?

Chemicals Which Affect Plant Growth: Their Use and Abuse

The subtle chemical balance maintained in nature between populations may be disturbed, for good or ill, by the widespread use and abuse of chemical growth regulators as herbicides and pesticides in agricultural practice (Orians and Pfeiffer, 1970; Report of the Secretary, 1969; Whiteside, 1970a,b).

Through the progressive development of the industrial age, man has been remarkably complacent about chemical pollutants which have had a dramatic effect on vegetation (see "Pesticides and Their Effects on Soil and Water," 1966; Hindawi, 1970; Jacobson and Hill, 1970). One only needs to visit the vicinity of large-scale smelting operations to see the man-made deserts which arise. But from SO_2 from smelting operations (Fig. 10-5), to HF from phosphate plants (Fig. 10-6), to mercury compounds from pulp mills, to carbon monoxide and the contaminants from coal gas, to the more subtle consequences of the by-products of the petroleum industry, represents a chain of events by which man has progressively polluted his environment and the one in which, perforce, plants live (see Table 10-1). All these compounds which are the products of an industrial society, whether obvious in large amounts or subtle in small amounts, have created for plants an increasingly hostile environment (Dugger and Ting, 1970a,b).

Yet the view has been taken by some that plants may in fact cleanse polluted air, even as they are being injured. It is well known that ozone is conspicuous in the city-engendered gasoline-type "smog." More specifically, peroxyacetyl nitrate (PAN)

$$CH_3—\overset{\overset{\displaystyle O}{\|}}{C}—OONO_2$$

and ozone are products of the photochemical reactions of the hydrocarbons and nitrogen oxides. Nitrogen dioxide absorbs ultraviolet light when exposed to sunlight and utilizes the energy to break the bond between oxygen and nitrogen. This reaction yields nitric oxide and atomic oxygen, which in turn cause a number of oxidations with ozone as the principal oxidant produced.

$$NO_2 + O_2 \underset{}{\overset{hv}{\rightleftarrows}} NO + O_3$$

The ozone thus formed can severely injure a number of plants (see Table 10-1 for symptoms). According to Waggoner and co-workers (see Rich *et al.*, 1970), ozone causes "leakage" from leaf cells to the extent that their surfaces may even become wet. Measurements indicate that ozone is re-

Fig. 10-5. Sulfur dioxide damage in conifers. Top, the normal habit of vigorous shoots of spruce (*Picea abies*); bottom, shoots of comparable plants from an area affected by sulfur dioxide from an aluminum smelting operation. (Photograph courtesy of Dr. Clarence Gordon, University of Montana, Missoula.)

duced to a very low concentration at surfaces of substomatal cells and that the removal of ozone from the air is "regulated by the same factors that control the exchange of water vapor between leaves and the atmosphere" (Rich *et al.*, 1970). Moreover, ozone is absorbed in the presence of light, and dramatically increases stomatal resistance in plants, such as tobacco and beans, thus emphasizing the view that for smog damage to occur stomata must be open (see Dugger and Ting, 1970a,b).

As a palliative to ozone injury in some plants, such as tobacco, some reducing agents work, but phenylmercuric acetate (PMA) is a corrective. This is presumably an indirect effect, as PMA induces stomatal closure (see Zelitch, 1969).

To what extent plants are able to "cleanse the air" is difficult to determine, as the plants must take up the pollutants and presumably disperse these substances in one way or another. Bernatzsky (1966) takes the view that stands of vegetation can be significant toward reducing particulate matter in polluted air because a lowered rate of air movement enables particles to settle on leaf surfaces. However, Bernatzsky is sceptical of any important effects by means of systemic cleansing. There is a serious need for concerted studies in this area.

Whereas the above considerations raise severe agricultural problems in terms of food production, conservation of natural resources, etc., there has been little concern for the hazardous indirect consequences on humans and animals of plant growth-regulating chemicals used agriculturally. This important area should not be overlooked.

Subsequent to the development of 2,4-D during World War II, the peacetime use of this and related substances in the control of weeds (as defoliants and in a variety of other ways as agricultural chemicals) became the major industrial enterprise which is now familiar (Frear, 1969; Menzie, 1966). Full credit should be given to the benefits achieved in agriculture by the controlled use of these substances, even at a time when their abuse is being widely deplored (see Galston, 1967a, 1970). (The substances that are most dangerous are obviously those which are not readily degraded (Freed, 1966; Kearney and Kaufman, 1969). The very fact that these substances, ranging from 2,4-D and NAA to a large number of synthetic and related compounds have long been known at low concentration to stimulate, synergistically, cell division should have sounded the word of warning. Recently, it has been shown that such substances (especially 2,4,5-T), either by themselves or by virtue of their impurities (such as dioxins) or reaction products, are teratogens, i.e., they intervene in embryo development to cause abnormalities (Anonymous, 1970; Courtney *et al.*, 1970; Drill and Hiratzka, 1953; Report of the Secretary, 1969). Thus, the use of

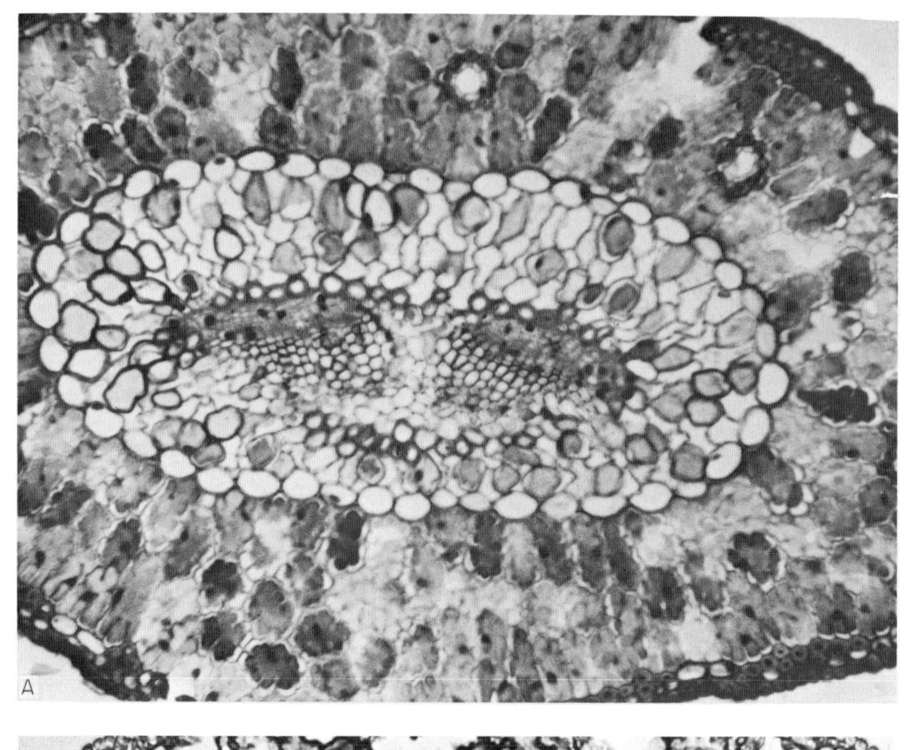
A

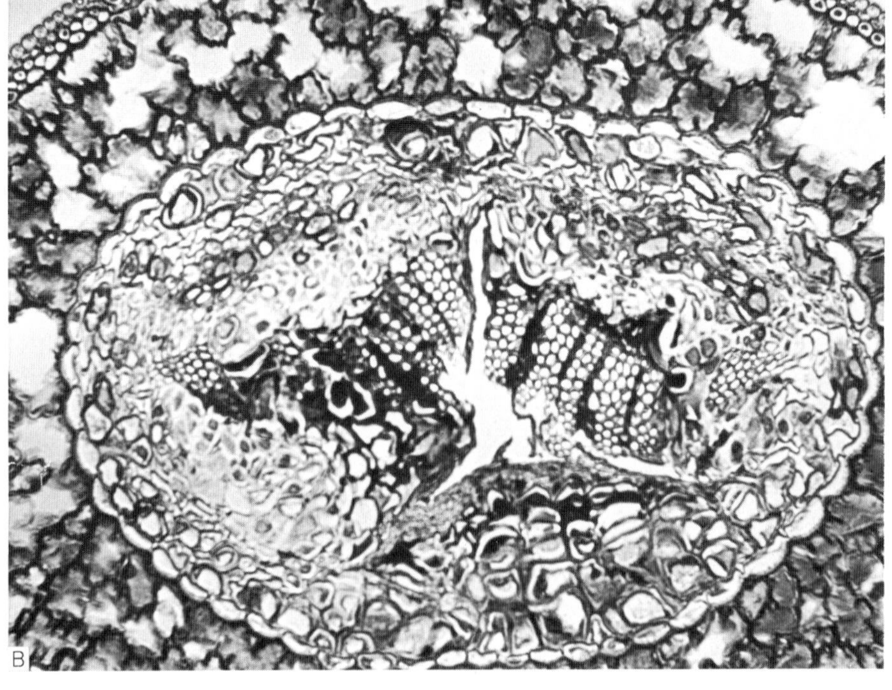
B

such substances in amounts or situations such that residues may enter or destroy the various food chains raises both questions of expediency and morality (Galston, 1967a, 1970; Mayer, 1967, Nelson, 1970; Report to the Subcommittee on Science, Research and Development, 1969). Even the occurrence of so-called secondary products in plants, such as alkaloids, cyanogenic glucosides, and phenolics, may be modified by the use of growth regulators and fertilizers in agriculture; this has had consequences that were not initially foreseen (Keeler, 1969; Singleton and Kratzer, 1969). From this point of view, the quantities of such chemicals that have been used in both peace and war are far in excess of reasonable limits, as recent investigation has emphasized (see Nelson, 1970; Report to the Subcommittee on Science, Research and Development, 1969; Whiteside, 1970a,b).

In all these contexts, therefore, research into the regulatory substances that control plant development becomes not an academic preoccupation but one which raises profound problems of human welfare. The imbalanced usage of almost any chemical substance, whether this is intrinsically beneficial or not (e.g., even NO_3 or PO_4), may disturb the ecosystem in undesirable ways. Among the unexpected effects of our chemical environment, an inadvertant combination of chemical agents could render organisms in key cycles (e.g., the nitrogen cycle) ineffective, thereby causing irreparable damage.

Even the screening tests which are widely practiced in industry to guide, empirically, the course of development of new compounds have their shortcomings. They commonly focus attention upon easily recognizable criteria of action, and they usually emphasize the role of compounds acting alone. But it is the indirect and unexpected responses that often give cause for concern, even as it is the interactions of the substances in question with others in the organism or the environment which may lead to either useful or unpredictable results. In this connection it is interesting to note that agents, such as α-hydroxysulfonates (e.g., α-hydroxy-2-pyridinemethanesulfonic acid), can inhibit stomatal opening in the presence of light and can close already open stomata (see Zelitch, 1969) by acting competetively as an inhibitor of glycolate oxidase. Zelitch has shown that α-hydroxy-2-pyridinemethanesulfonic acid specifically inhibits photorespiration, with the resultant accumulation of glycolate, without inhibiting dark respiration. Photorespiration, a special kind of respiration which occurs only during

---

Fig. 10-6. Hydrogen fluoride damage in conifers. A, a transverse section of normal needles of ponderosa pine (*Pinus ponderosa*); B, a similar section showing the damage caused by hydrogen fluoride from an industrial operation. (Photograph courtesy of Dr. Clarence Gordon, University of Montana, Missoula.)

TABLE 10-1

SUMMARY OF ATMOSPHERIC POLLUTANTS, SOURCES, SYMPTOMS, VEGETATION AFFECTED, INJURY THRESHOLDS, AND CHEMICAL ANALYSES[a]

Pollutants	Source	Symptom	Type of leaf affected	Part of leaf affected	Injury threshold[b]			Chemical analysis for pollutants in plants
					ppm	μg/m³	Sustained exposure	
Ozone (O_3)	Photochemical reaction of hydrocarbon and nitrogen oxides from fuel combustion, refuse burning, and evaporation from petroleum products and organic solvents	Fleck, stipple, bleaching, bleached spotting, pigmentation, growth suppression, and early abscission. Tips of conifer needles become brown and necrotic	Old, progressing to young	Palisade cells	0.03	70	4 hr	None
Peroxyacetyl nitrate (PAN)	Same sources as ozone	Glazing, silvering, or bronzing on lower surface leaves	Young	Spongy cells	0.01	250	6 hr	None
Nitrogen dioxide (NO_2)	High-temperature combustion of coal, oil, gas, and gasoline in power plants and internal combustion engines	Irregular, white, or brown collapsed lesion on intercostal tissue and near leaf margin	Middle-aged	Mesophyll cells	2.5	4700	4 hr	None
Sulfur dioxide (SO_2)	Coal, fuel oil, and petroleum	Bleached spots, bleached areas between veins, bleached margin, chlorosis, growth suppression, early abscission, and reduction in yield	Middle-aged	Mesophyll cells	0.3	800	8 hr	[c]

Hydrogen fluoride (HF)	Phosphate rock processing, aluminum industry, iron smelting, brick and ceramic works, and fiber glass manufacturing	Tip and margin burn, chlorosis, dwarfing, leaf abscission, and lower yield	Mature	Epidermis and mesophyll	0.1 (ppb)	0.2	5 weeks	Distillation and titration
Chlorine (Cl_2)	Leaks in chlorine storage tanks; hydrochloric acid mist	Bleaching between veins, tip and margin burn, and leaf abscission	Mature	Epidermis and mesophyll	0.10	300	2 hr	[c]
Ethylene ($CH_2{=}CH_2$)	Incomplete combustion of coal, gas, and oil for heating, and automobile and truck exhaust	Sepal withering, leaf abnormalities; flower dropping, and failure of flower to open properly	Flower	All	0.05	60	6 hr	None

[a] From Hindawi, 1970.
[b] Metric equivalent based on 25°C and 760 mm mercury.
[c] Chemical analysis often is not reliable for diagnosing chloride or sulfate accumulation in leaf tissue because undamaged plants often contain higher concentrations of these pollutants than are found in damaged plants.

photosynthesis and is distinct from dark aerobic respiration, consumes glycolic acid (H_2C—OH—COOH) as its substrate in organelles called peroxosomes. However, glycolic acid is produced by and accumulates in chloroplasts.

Glycolic acid is supposed to be a common intermediary between photosynthesis and photorespiration even though the sources of the carbon skeleton of glycolic acid are not wholly known (see Gibbs, 1970, and references cited therein). Plants that synthesize and metabolize glycolic acid (e.g., tobacco) are less efficient in terms of carbon assimilation per unit of light energy, consumed to the extent that the glycolic acid oxidizes to CO_2. The so-called "maize-type" plants (e.g., *Zea,* sugar cane, and sorghum), however, are very efficient, as they do not undergo glycolic acid metabolism and produce more dry weight per unit of water transpired.

The point of interest which derives from work of this kind is that α-hydroxy-2-pyridinemethanesulfonic acid "converts" the metabolism of isolated tobacco leaf discs (which normally have a high photorespiratory rate) to a "maize-type" plant with reduced photorespiration and, hence, an enhanced net photosynthesis.

Similarly, plants wilting in the light from temperature and wind effects may increase their abscisic acid content some 40-fold (see Wright and Hiron, 1969). There is an inferred causal relationship to stomatal response (closure) and through this a control of water content, since by closure plants can "recover" from wilting. The ABA increase is caused by synthesis from mevalonic acid (Milborrow, 1970). Analogs of ABA, 5-(1,2-epoxy-2,6,6-trimethyl-1-cyclohexyl-)-3-methyl-2,4-*cis, trans*-pentadienoic acid, are converted to ABA in large quantities, and wilting affects the conversion rate. Since the presence of the "epoxide" causes a reduced incorporation of H^3-mevalonic acid into ABA, it has been suggested that the epoxide or a close derivative is in the normal biosynthetic pathway to ABA. The discovery that ABA causes wilting has thus initiated a study of the complicated pathways leading to its biosynthesis (see Milborrow, 1970). Moreover, the accumulation of ABA in plants during wilting (Wright, 1969) has obvious implications for horticultural practices such as irrigation and seedling transplantation (Wright and Hiron, 1970).

All this emphasizes the pressing need to appreciate the important role that plant physiology and biochemistry should play. This should be so in conjunction with plant breeding programs, or in the technological development of new compounds and in relation to industrial processes which have effects upon the composition and behavior of plants. The relations to the physiology of plants should, in these areas, be a prime consideration, not an afterthought.

Summation

At each level of complexity and at each level of observation of plant response there are different degrees of freedom, which chemical growth regulators may exploit. These may range from effects upon the activity of a gene-determined enzyme system, which may be demonstrable *in vitro*, to the growth of isolated cells and the stimulus to their morphogenesis into plants; they may range from effects observed on subcellular organelles *in situ* to the complexity of the developmental responses of plants to their environment as they grow and reproduce. A still different level of growth regulation is observed in the proliferating cells of a legume nodule, for this is a feature of the interaction of the bacterium and the host. With selected strains of bacteria the cells of the nodule produce a biochemical milieu for nitrogen fixation, which is far more effective than that which can be duplicated or maintained *in vitro,* even with present knowledge of the enzymology of the fixation process (see Hardy and Knight, 1968).

The evident range of plant responses to exogenous chemicals is the measure of the difficulty in their causal interpretation. The problems of growth and development are not disposed of by the accepted knowledge that they proceed within the limits of a plan that is inherited, nor are they solved by knowing that, within this plan, a given chemical reaction may be rendered feasible by the presence and action of a given gene. Similarly, it is at the level of chemical reactions that the intervention of exogenous substances may seem most explicit, but a wide gap remains to be bridged. Reactions produce only substances; growth produces form—it generates complexity and produces organization.

All the above is not said in a defeatist frame of mind. The first step toward solving a problem is to understand its scope. If, to make problems amenable to current concepts or fashionable techniques, they are oversimplified—even misrepresented—it is salutary to correct this. The emphasis here, therefore, is frankly on a holistic approach to the responses of cells, organs, and organisms to extraneous chemical agents. This is forced upon one by the recognition that living cells of angiosperms are totipotent throughout, and they are constantly being regulated in what they do by properties of their immediate environment; moreover, they are also constituted as an array of autonomous organelles, which present a multiplicity of reactive sites, but they are nevertheless subject to overall control and direction.

The impressive facts of angiosperm development, seen in the study of growing points, present the challenge and also illustrate the philosophy of growth regulation. All the evidence indicates that the individual cells of shoot and root apices are genetically interchangeable, since, when isolated, they are potentially able to grow like zygotes. The same applies to many other living parenchyma, which nevertheless exhibit marked morphological and biochemical differentiation even in the adjacent cells of the same complex tissue system (Steward, 1970). Whether judged by the cell and tissue forms to which they give rise, or the biochemistry of their storage products *in situ*, it is clear that the cells react to "where they are" in the plant body to exhibit a characteristic behavior. Carrot cells from the storage root in free culture will turn green and produce chloroplasts; they will even give rise to embryos which form shoot and roots and later whole plants; but as yet they only form rich orange-red carotene bodies when they develop at the right time and place in the seedling axis (Israel *et al.*, 1969). When the subterminal region of a vegetative shoot, which contains relatively quiescent cells, is stimulated (e.g., by length of day, by night temperature, or other stimulus) to develop and give rise to flowers, or an influorescence, the impression is inescapable that the cells in question must receive some form of message via a "chemical messenger," and, in this case, its first effect is to stimulate them into activity and growth. The ability of living cells in the plant body to respond in all these ways is obviously inherent in their genetic constitution, but it is customary to describe their several responses during development as "epigenetic." Epigenesis is, however, only a name for the unknown factors of the immediate environment of cells which "tells them what to do where and when they are." In translating these questions into meaningful chemical and physical terms, a place must surely be found for the effects transmitted to cells which are mediated by a range of substances (i.e., growth-regulating substances) acting upon them singly, synergistically, and sequentially, as part of that complex or matrix of external factors that, from without, controls the expression *in situ,* or in culture, of their innate potentialities. The evidence presented here points to an ample array of substances and effects that may illustrate how this may be done. Furthermore, as cells respond in this way to chemical stimuli they may not need, throughout their life, continually and always to "ask the permission of their nuclei." The impression is that the responses of cells to exogenous substances may be reactions to perturbations in the whole matrix or network which was established *de novo* under the direction of the nucleus.

While molecular biology, as such, goes far to describe how given macromolecules are made, it has as yet contributed little toward our understanding of development and morphogenesis; in fact, the latter events are often necessary to provide the milieu for the former to function. As attached adjacent cells, or organs, embark on their often contrasted developmental sequences, some form of "molecular conversation" must be their internal channel of "communication." The varied exogenous growth-regulating substances seem to intervene to modify the consequences of this intercellular communication, and their effects may become manifest at any of the various levels of organization, i.e., in organelles, in cells, organs, or organisms. In the final analysis, it is cells—not enzymes, not genes, not metabolites—that grow, and the plant growth regulators become accessible to, and exercise controls over, any of the events which by their very complexity, balance, and coordination distinguish the "quick from the dead." Nineteenth century writers assigned to living systems an essential ability to react to stimuli, which they termed "irritability," and, despite our modern biochemical knowledge, this outmoded term still covers much that is not understood. Thus, a full and final knowledge of plant growth regulators, intimately related to knowledge of fine structure and morphology, could go far toward penetrating the essential mystery of the life of plants.

Bibliography

Addicott, F. T. (1968). Environmental factors in the physiology of abscission. *Plant Physiol.* **43**, 1471–1479.

Addicott, F. T., and Lyon, J. L. (1969). Physiology of abscisic acid and related substances. *Annu. Rev. Plant Physiol.* **20**, 139–164.

Addicott, F. T., Carns, H. R., Lyon, J. L., Smith, O. E., and McMeans, J. L. (1964). On the physiology of abscisins. *In* "Régulateurs naturels de la croissance végétale" (J. P. Nitsch, ed.), pp. 687–703. C. N. R. S., Paris.

Addicott, F. T., Lyon, J. L. Ohkuma, K., Thiessen, W. E., Carns, H. R., Smith, O. E., Cornforth, J. W., Milborrow, B. V., Ryback, G., and Wareing, P. F. (1968). Abscisic acid: A new name for Abscisin II (Dormin). *Science* **159**, 1493.

Adelmann, H. B. (1966). "Marcello Malpighi and the Evolution of Embryology," 5 vols. Cornell Univ. Press, Ithaca, New York.

Ahmadjian, V. (1967). "The Lichen Symbiosis." Ginn (Blaisdell), Boston, Massachusetts.

Allende, J. E. (1969). Protein biosynthesis in plant systems. *Tech. Protein Biosyn.* **2**, 55–100.

Ammirato, P. V., and Steward, F. C. (1969). Indirect effects of irradiation: Morphogenetic effects of irradiated sucrose. *Develop. Biol.* **19**, 87–106.

Anderson, G. R., Jenner, E. L., and Mumford, F. E. (1969). Temperature and pH studies on phytochrome *in vitro*. *Biochemistry* **8**, 1182–1187.

Anderson, L., and Wolter, K. E. (1966). Cyclitols in plants: Biochemistry and physiology. *Annu. Rev. Plant Physiol.* **17**, 209–222.

Anonymous (1970). Defoliants, deformities: What risk? *Med. World News* **11**, [9], 15–17.

Armstrong, D. J., Burrows, W. J., Skoog, F., Roy, K. L., and Soll, D. (1969). Cytokinins: Distribution in transfer RNA species of *Escherichia coli*. *Proc. Nat. Acad. Sci. U. S.* **63**, 834–841.

Arthur, J. M., and Harvill, E. K. (1941/2). Flowering in *Digitalis purpurea* initiated by low temperature and light. *Contrib. Boyce Thompson Inst.* **12**, 111–117.

Audus, L. J. (1963). "Plant Growth Substances." Leonard Hill, London.

Audus, L. J., ed. (1964). "The Physiology and Biochemistry of Herbicides." Academic Press, New York.

Audus, L. J. (1969). Geotropism. *In* "The Physiology of Plant Growth and Development" (M. B. Wilkins, ed.), pp. 205–242. McGraw-Hill, New York.

Avery, G. S., Jr. (1930). Comparative anatomy and morphology of embryos and seedlings of maize, oats, and wheat. *Bot. Gaz.* **89**, 1–39.

Bachi, M. D., Epstein, J. W., Herzberg-Minzly, Y., and Loewenthal, H. J. E. (1969). Synthesis of compounds related to gibberellic acid. III. Analogs of ring A of the gibberellins. *J. Org. Chem.* **34**, 126–135.

Baldev, B., Lang, A., and Agatep, A. O. (1965). Gibberellin production in pea seeds developing in excised pods: Effect of growth retardant AMO-1618. *Science* **147**, 155–157.

Ball, E. (1946). Development in sterile culture of stem tips and subjacent regions of *Tropaeolum majus* L. and *Lupinus albus* L. *Amer. J. Bot.* **33**, 301–318.

Ball, N. G. (1969). Tropic, nastic, and tactic responses *In* "Plant Physiology: A Treatise" (F. C. Steward, ed.), Vol. 5A, pp. 119–228. Academic Press, New York.

Baltimore, D. (1970). RNA-dependent DNA polymerase in virions of RNA tumour viruses. *Nature (London)* **226**, 1209–1211.

Bandurski, R. S., Ueda, M., and Nicholls, P. B. (1969). Esters of indole-3-acetic acid and *myo*-inositol. *Ann. N. Y. Acad. Sci.* **165**, 655–667.

Barber, J. T., and Steward, F. C. (1968). The proteins of *Tulipa* and their relation to morphogenesis. *Develop. Biol.* **17**, 326–349.

Barber, J. T., Srb, A. M., and Steward, F. C. (1969). Proteins, morphology and genetics in *Neurospora. Develop. Biol.* **20**, 105–124.

Barker, W. G., and Steward, F. C. (1962). Growth and development of the banana plant. I. The growing regions of the vegetative shoot. *Ann. Bot. (London)* [N. S.] **26**, 389–411.

Barth, A., and Mitchell, H.-J. (1969). Wirkungsmechanismen und Substituenteneinflüsse von Herbiziden des Harnstoff-, Carbamat-, Amid- und Triazintyps. *Pharmazie* **24**, 11–23.

Bartz, J., Söll, D., Burrows, W. J., and Skoog, F. (1970). Identification of the cytokinin-active ribonucleosides in pure *Escherichia coli* tRNA species. *Proc. Nat. Acad. Sci. U.S.* **67**, 1448–1453.

Beckwith, J. R., and Zisper, D., eds. (1970). "The Lactose Operon," Cold Spring Harbor Monogr. Cold Spring Harbor, New York.

Beermann, W. (1966). Differentiation at the level of the chromosomes. *In* "Cell Differentiation and Morphogenesis," pp. 24–54. North Holland Publ. Amsterdam and Wiley, New York.

Bellamy, A. R. (1966). RNA synthesis in exponentially growing tobacco cells subjected to a step-down nutritional shift. *Biochim. Biophys. Acta* **123**, 102–115.

Bentley, J. A. (1958). The naturally occurring auxins and inhibitors. *Annu. Rev. Plant. Physiol.* **9**, 47–80.

Bentley, J. A. (1961). Chemistry of the native auxins. *In* "Handbuch der Pflanzenphysiologie" (W. Ruhland, ed.), Vol. XIV, pp. 485–500. Springer, Berlin.

Bergmann, L. (1967). Wachstum grüner suspensionskulteren von *Nicotiana tabacum* var. "Samsun" mit CO_2 als Kohlenstoffquelle. *Planta* **74**, 243–249.

Bergmann, L. (1968). Photosynthesis and growth of suspension cultures of *Nicotiana tabacum* with CO_2 as carbon source. *In* "Les Cultures de Tissus de Plantes, (R. J. Gautheret and L. Hirth, eds.), pp. 213–221. C. N. R. S., Paris.

Bernatzsky, A. (1966). Climatic influence of the greens and city planning. *Anthos* **1**, 29–36.

Bernier, G., ed. (1970). "Cellular and Molecular Aspects of Floral Induction." Longman Group, London.

Berridge, M. V., Ralph, R. K., and Letham, D. S. (1970). The binding of kinetin to plant ribosomes. *Biochem. J.* **119**, 75–84.

Berrie, A. M. M., Parker, W., Knights, B. A., and Hendrie, M. R. (1968). Studies on lettuce seed germination. I. Coumarin induced dormancy. *Phytochemistry* **7**, 567–573.

Birnstiel, M. (1967). The nucleolus in cell metabolism. *Annu. Rev. Plant Physiol.* **18**, 25–58.

Black, M. (1969). Light-controlled germination of seeds. *Symp. Soc. Exp. Biol.* **23**, 193–217.

Blackman, F. F. (1905). Optima and limiting factors. *Ann. Bot. (London)* **19**, 281–295.
Bloch, R. (1941). Wound healing in higher plants. *Bot. Rev.* **7**, 110–146.
Bloch, R. (1952). Wound healing in plants. *Bot. Rev.* **18**, 655–679.
Boardman, N. K. (1968). The photochemical systems of photosynthesis. *Advan. Enzymol.* **30**, 1–79.
Bock, R. M. (1970). Nucleic acid structure function relations. *Science* **170**, 351–355.
Bogorad, L. (1967). The role of cytoplasmic units. Control mechanisms in plastid development. *In* "Control Mechanisms in Developmental Processes" (M. Locke, ed.), pp. 1–31. Academic Press, New York.
Boisard, J., and Malcoste, R. (1970). Analyse spectrophotométrique du phytochrome dans l'embryon de Courge (*Cucurbita pepo*) et de Potiron (*Cucurbita maxima*). *Planta* **91**, 54–67.
Bonner, J. (1965). "The Molecular Biology of Development." Oxford Univ. Press, London and New York.
Bonner, J., and English, J., Jr. (1938). A chemical study of traumatin, a plant wound hormone. *Plant Physiol.* **13**, 331–348.
Bonner, J., and Liverman, J. (1953). Hormonal control of flower initiation. *In* "Growth and Differentiation in Plants" (W. E. Loomis, ed.), pp. 283–303. Iowa State Coll. Press, Ames, Iowa.
Bonner, J., Heftmann, E., and Zeevaart, J. (1963). Suppression of floral induction by inhibitors of steroid biosynthesis. *Plant Physiol.* **38**, 81–88.
Bonner, J., Dahmus, M. E., Fambrough, D., Huang, R. -c., Marushige, K., and Tuan, D. Y. H. (1968). The biology of isolated chromatin. *Science* **159**, 47–56.
Bonnet, C. (1754). "Recherches sur l'usage des feuilles dans les plantes, et sur quelques autres sujets relativs à l'histoire de la végétation." Elie Luzac, Fils. Imp. -Lib., Gottingue & Leide.
Bonnet, O. T. (1961). The oat plant: Its histology and development. *Ill., Agr. Exp. Sta., Bull.* **672**, 1–112.
Börner, H. (1960). Liberation of organic substances from higher plants and their role in the soil sickness problem. *Bot. Rev.* **26**, 393–424.
Boulter, D. (1970). Protein synthesis in plants. *Annu. Rev. Plant Physiol.* **21**, 91–114.
Boysen Jensen, P. (1936). "Growth Hormones in Plants" (transl. by G. S. Avery and P. R. Burkholder). McGraw-Hill, New York.
Brachet, J. (1957). "Biochemical Cytology." Academic Press, New York.
Brachet, J. (1968). Synthesis of macromolecules and morphogenesis in *Acetabularia*. *Curr. Top. Develop. Biol.* **3**, 1–36.
Brandes, H., and Kende, H. (1968). Studies on cytokinin-controlled bud formation in moss protonemata. *Plant Physiol.* **43**, 827–837.
Braun, A. C. (1961). The origin of the plant tumor cell. *In* "Growth in Living Systems" (M. X. Zarrow, ed.), pp. 605–619. Basic Books, New York.
Braun, A. C. (1969a). Abnormal growth in plants. *In* "Plant Physiology: A Treatise" (F. C. Steward, ed.), Vol. 5B, pp. 379–420. Academic Press, New York.
Braun, A. C. (1969b). "The Cancer Problem." Columbia Univ. Press, New York.
Braun, A. C. (1970). On the origin of the cancer cells. *Amer. Sci.* **58**, 307–320.
Braun, A. C., and Stonier, T. (1958). Morphology and physiology of plant tumors. *In* "Protoplasmatologia, Handbuch der Protoplasmaforschung" (L. V. Heilbrunn and F. Weber, eds.), Vol. X, Part 5a, pp. 1–93. Springer, Vienna.
Braun, A. C., and Wood, H. N. (1969). Transformation and recovery of the crown gall tumor cell: An experimental model. *Fed. Proc., Fed. Amer. Soc. Exp. Biol.* **28**, 1815–1819.
Brian, P. W. (1957). Effects of antibiotics on plants. *Annu. Rev. Plant Physiol.* **8**, 413–426.
Brian, P. W. (1966). The gibberellins as hormones. *Int. Rev. Cytol.* **19**, 229–266.

Brian, P. W., and Hemming, H. G. (1955). The effect of gibberellic acid on shoot growth of pea seedlings. *Physiol. Plant.* **8**, 669–681.

Brian, P. W., Grove, J. F., and MacMillan, J. (1960). The Gibberellins. *Fortschr. Chem. Org. Naturst.* **18**, 350–433.

Brian, P. W., Grove, J. F., and Mulholland, T. P. C. (1967). Relationships between structure and growth-promoting activity of the gibberellins and some allied compounds in four test systems. *Phytochemistry* **6**, 1475–1499.

Brinks, R., MacMillan, J., and Pryce, R. J. (1969). Plant hormones. VIII. Combined gas chromatography-mass spectrometry of the methyl esters of gibberellins A_1 to A_{24} and their trimethylsilylethers. *Phytochemistry* **8**, 271–284.

Britten, R. J., and Davidson, E. H. (1969). Gene regulation for higher cells: a theory. *Science* **165**, 349–357.

Brookes, R. F., and Leafe, E. L. (1963). Structure and plant growth regulating activity of 2-benzothiazolyloxyacetic acids and 2-oxobenzothiazolin-3-ylacetic acids. *Nature (London)* **198**, 589.

Bruce, M. I., and Zwar, J. A. (1966). Cytokinin activity of some substituted ureas and thioureas. *Proc. Roy. Soc., Ser. B* **165**, 245–265.

Bücher, T., and Sies, H., eds. (1969). "Inhibitors, Tools in Cell Research." Springer, New York.

Bünning, E. (1967). "The Physiological Clock." Springer, Berlin.

Burg, S. P., and Burg, E. A. (1968). Auxin stimulated ethylene formation: Its relationship to auxin inhibited growth, root geotropism and other plant processes. *In* "Biochemistry and Physiology of Plant Growth Substances" (F. Wightman and G. Setterfield, eds.), pp. 1275–1294. Runge Press, Ottawa.

Burrows, W. J., Armstrong, D. J., Skoog, F., Hecht, S. M., Boyle, J. T. A., Leonard, N. J., and Occolowitz, J. (1969). The isolation and identification of two cytokinins from *Escherichia coli* transfer ribonucleic acids. *Biochemistry* **8**, 3071–3076.

Burrows, W. J., Armstrong, D. J., Kaminek, M., Skoog, F., Bock, R. M., Hecht, S. M., Damman, L. G., Leonard, N. J., and Occolowitz, J. (1970). Isolation and identification of four cytokinins from wheat germ transfer ribonucleic acids. *Biochemistry* **9**, 1867–1872.

Caplin, S. M., and Steward, F. C. (1948). Effect of coconut milk on the growth of explants from carrot root. *Science* **108**, 655–657.

Caplin, S. M., and Steward, F. C. (1952). Investigations on the growth and metabolism of plant cells. II. Variables affecting the growth of tissue explants and the development of a quantitative method using carrot root. *Ann. Bot. (London)* [N. S.] **16**, 219–234.

Carr, D. J., and Reid, D. M. (1968). The physiological significance of hormones in roots and their export to the shoot system. *In* "Biochemistry and Physiology of Plant Growth Substances" (F. Wightman and G. Setterfield, eds.), pp. 1169–1185. Runge Press, Ottawa.

Cathey, H. M. (1964). Physiology of growth retarding chemicals. *Annu. Rev. Plant Physiol.* **15**, 271–302.

Cathey, H. M., and Steffens, G. L. (1968). Relation of the structure of fatty acid derivatives to their action as chemical pruning agents. *In* "Plant Growth Regulators," SCI Monogr. No. 31, pp. 224–235.

Cathey, H. M., Steffens, G. L., Stuart, N. W., and Zimmerman, R. H. (1966). Chemical pruning of plants. *Science* **153**, 1382–1383.

Chailakhyan, M. Kh. (1937). Concerning the hormonal nature of plant development processes. *Dokl. Acad. Sci. USSR* **16**, 227–230; translated and reprinted, *in* "Papers on Plant Growth and Development" (W. M. Laetsch and R. E. Cleland, eds.), pp. 364–369. Little, Brown, Boston, Massachusetts, 1967.

Chailakhyan, M. Kh. (1961). Principles of ontogenesis and physiology of flowering in higher plants. *Can. J. Bot.* **39**, 1817–1841.

Chailakhyan, M. Kh. (1964). Florigen, gibberellins, and anthesins. *In* "Régulateurs naturels de la croissance végétale" (J. P. Nitsch, ed.), pp. 589–596. C. N. R. S., Paris.

Chailakhyan, M. Kh. (1968a). Flowering hormones of plants. *In* "Biochemistry and Physiology of Plant Growth Substances" (F. Wightman and G. Setterfield, ed.), pp. 1317–1340. Runge Press, Ottawa.

Chailakhyan, M. Kh. (1968b). Internal factors of plant flowering. *Annu. Rev. Plant Physiol.* **19**, 1–36.

Cherry, J. H. (1970). Nucleic acid synthesis during floral induction of *Xanthium*. *In* "Cellular and Molecular Aspects of Floral Induction." (G. Bernier, ed.), pp. 173–191. Longman Group, London.

Cherry, J. H., and van Huystee, R. B. (1965). Comparison of messenger RNA in photoperiodically induced and noninduced *Xanthium* buds. *Science* **150**, 1450–1453.

Chibnall, A. C. (1939). "Protein Metabolism in the Plant." Yale Univ. Press, New Haven, Connecticut.

Chouard, P. (1960). Vernalization and its relations to dormancy. *Annu. Rev. Plant Physiol.* **11**, 191–237.

Cleland, R. (1965). Auxin-induced cell wall loosening in the presence of actinomycin D. *Plant Physiol.* **40**, 595–600.

Cleland, R. (1968). Auxin and wall extensibility: Reversibility of auxin-induced wall-loosening process. *Science* **60**, 192–194.

Cleland, R., Thompson, M. L., Rayle, D. L., and Purves, W. K. (1968). Difference in effects of gibberellins and auxins on wall extensibility of cucumber hypocotyls. *Nature (London)* **219**, 510–511.

Clever, U., and Ellgaard, E. G. (1970). Puffing and histone acetylation in polytene chromosomes. *Science* **169**, 373–374.

Clowes, F. A. L. (1961). "Apical Meristems." Davis, Philadelphia, Pennsylvania.

Clowes, F. A. L., and Juniper, B. E. (1968). "Plant Cells," Bot. Monogr., Vol. 8. Blackwell, Oxford.

Côme, D. (1970). "Les Obstacles à la germination." Masson, Paris.

Commoner, B. (1968). Failure of the Watson-Crick theory as a chemical explanation of inheritance. *Nature* (*London*) *220*, 334–340.

Conn, E. E., and Butter, G. W. (1969). The biosynthesis of cyanogenic glycosides and other simple nitrogen compounds. *In* "Perspectives in Phytochemistry" (J. B. Harborne and T. Swain, eds.), Academic Press, New York.

Cooke, A. R., and Randall, D. I. (1968). 2-haloethane phosphonic acids as ethylene releasing agents for the induction of flowering in pineapples. *Nature (London)* **218**, 974.

Coombe, B. G., Cohen, D., and Paleg, L. G. (1967). Barley endosperm bioassay for gibberellins. II. Application of the method. *Plant Physiol.* **42**, 113–119.

Cornforth, J. W., Milborrow, B. V., Ryback, G., Rothwell, K., and Wain, R. L. (1966). Identification of the yellow lupin growth inhibitor as (+)-Abscisin II ((+)-dormin). *Nature (London)* **211**, 742–743.

Correll, D. L., Edwards, J. L., and Shropshire, W., Jr. (1968). Multiple chromophore species in phytochrome. *Photochem. Photobiol.* **8**, 465–475.

Courtney, K. D., Gaylor, D. W., Hogan, M. C., Falk, H. L., Bates, R. R., and Mitchell, I. (1970). Teratogenic evaluation of 2,4-5 T. *Science* **168**, 864–866.

Cox, R. A. (1968). Macromolecular structure and properties of ribonucleic acids. *Quart. Rev., Chem. Soc.* **22**, 499–526.

Cross, B. E. (1968). Biosynthesis of gibberellins. *Prog. Phytochem.* **1**, 195–222.

Crozier, A., Kuo, C. C., Durley, R. C., and Pharis, R. P. (1970). The biological activities of 26 gibberellins in nine plant bioassays. *Can. J. Bot.* **48**, 867–877.

Cumming, B. G., and Wagner, E. (1968). Rhythmic processes in plants. *Annu. Rev. Plant Physiol.* **19**, 381–416.

Curry, G. M. (1969). Phototropism. *In* "The Physiology of Plant Growth and Development" (M. B. Wilkins, ed.), pp. 245–273. McGraw-Hill, New York.

Curtis, O. F., and Chang, H. T. (1930). The relative effectiveness of the temperature of the crown as contrasted with that of the rest of the plant upon the flowering of celery. *Amer. J. Bot.* **17**, 1047–1048.

Cutter, E. G. (1965). Recent experimental studies of the shoot apex and shoot morphogenesis. *Bot. Rev.* **31**, 7–113.

Dahlhelm, H. (1969). Untersuchungen zum Wirkungsmechanismus von Indolyl-3-Essigsäure mit Hilfe von schwerem Wasser. *Planta* **86**, 224–234.

Dalton, L. K., and Brown, B. T. (1970). A new group of compounds with morphactin-like activity in *Arabidopsis thaliana. Seventh International Conference on Plant Growth Substances.* No. 20, *December 7–12* (Abstr.). *Canberra, Australia.*

D'Amato, F. (1952). Polyploidy in the differentiation and function of tissues and cells in plants. *Caryologia* **4**, 311–358.

Darlington, A. (1968). "The Pocket Encyclopedia of Plant Galls" (with illustrations by M. J. D. Hirons). Blandford Press, London.

Darwin, C., and Darwin, F. (1880). "The Power of Movement in Plants." John Murray, London.

Darwin, C., and Darwin, F. (1966). "The Power of Movement in Plants" (with a preface by Barbara Gillespie Pickard). De Capo Press, New York.

Datko, A. H., and Maclachlan, G. A. (1968). Indoleacetic acid and the synthesis of glucanases and pectic enzymes. *Plant Physiol.* **43**, 735–742.

Davidson, E. H. (1969). "Gene Activity in Early Development." Academic Press, New York.

Davis, B. D. (1968). Is riboflavin the photoreceptor in the induction of two-dimensional growth in fern gametophytes? *Plant Physiol.* **43**, 1165–1167.

Degani, N., and Steward, F. C. (1968). The effect of various media on the growth responses of different clones of carrot explants. *Ann. Bot.* (*London*) [N. S.] **32**, 97–117.

Dela Fuente, R. K., and Leopold, A. C. (1970). Time course of auxin stimulations of growth. *Plant Physiol.* **46**, 186–189.

DeMaggio, A. E. (1966). Phloem differentiation: Induced stimulation by gibberellic acid. *Science* **152**, 370–372.

Despois, R. (1958). Coumarins and phenolic acids in relation to inhibition and activation of germination. *In* "Biochemistry of Antibiotics" (K. Y. Spitzy and R. Brunner, eds.), pp. 33–36. Pergamon Press, Oxford.

Deuel, P. G., and Geissman, T. A. (1957). Xanthinin. II. The structures of Xanthinin and Xanthatin. *J. Amer. Chem. Soc.* **79**, 3778–3783.

Deysson, G. (1968). Antimitotic substances. *Int. Rev. Cytol.* **24**, 99–148.

Dimond, A. E. (1959). Some biochemical aspects of disease in plants. *Fortschr. Chem. Org. Natursto.* **17**, 298–321.

Dixon, R. O. D. (1969). Rhizobia. *Annu. Rev. Microbiol.* **23**, 137–158.

Drill, V. A., and Hiratzka, T. (1953). Toxicity of 2,4-dichlorophenoxyacetic acid and 2,4,5-trichlorophenoxyacetic acid. *Arch. Ind. Hyg.* **7**, 61–67.

Dudock, B. S., Katz, G., Taylor, E. K., and Holley, R. W. (1969). Primary structure of wheat germ phenylalanine transfer RNA. *Proc. Nat. Acad. Sci. U.S.* **62**, 941–945.

Duffus, C. M., and Duffus, J. H. (1969). A possible role cyclic AMP in Gibberellic acid triggered release of α-amylase in barley endosperm slices. *Experientia* **25**, 54.

Dugger, W. M., and Ting, I. P. (1970a). Air pollution oxidants—their effects on metabolic processes in plants. *Annu. Rev. Plant Physiol.* **21**, 215–234.

Dugger, W. M., and I. P. Ting (1970b). Physiological and biochemical effects of air pollution oxidants on plants. *Recent Advan. Phytochem.* **3**, 31–58.

Dunn, D. B., and Hall, R. H. (1970). Purines, pyrimidines, nucleosides and nucleotides: Physical constant and spectral properties. *In* "Handbook of Biochemistry," G-3 et seq., Chem. Rubber Publ. Co., Cleveland, Ohio.

Duysens, L. N. M., and Amesz, J. (1962). Function and identification of two photochemical systems in photosynthesis. *Biochim. Biophys. Acta* **64**, 243–260.

Dyer, T. A., and Leech, R. M. (1968). Chloroplast and cytoplasmic low-molecular-weight ribonucleic acid components of the leaf of *Vicia faba* L. *Biochem. J.* **106**, 689–698.

Dyson, W. H., Chen, C. M., Alam, S. N., Hall, R. H., Hong, C. I., and Chheda, G. B. (1970). Cytokinin activity of ureidopurine derivatives related to a modified nucleoside found in transfer RNA. *Science* **170**, 328–330.

Dzhaparidze, L. I. (1967). "Sex in Plants" (translated from Russian). Israel Program Sci. Transl., Jerusalem.

El-Antably, H. M. M., Wareing, P. F., and Hillman, J. (1967). Some physiological responses to d,l abscisin (dormin). *Planta* **73**, 74–90.

Ellis, R. J., and MacDonald, I. R. (1970). Specificity of cycloheximide in plants. *Plant Physiol.* **46**, 227–232.

Emmer, M., de Crombrugghe, B., Pastan, I., and Perlman, R. (1970). Cyclic AMP receptor protein of *E. coli*: Its role in the synthesis of inducible enzymes. *Proc. Nat. Acad. Sci. U.S.* **66**, 480–487.

English, J., Jr., Bonner, J., and Haagen-Smit, A. J. (1939a). The wound hormones of plants. II. The isolation of a crystalline active substance. *Proc. Nat. Acad. Sci. U.S.* **25**, 323–329.

English, J., Jr., Bonner, J., and Haagen-Smit, A. J. (1939b). The wound hormones of plants. IV. Structure and synthesis of a traumatin. *J. Amer. Chem. Soc.* **61**, 3434–3436.

Epstein, E., Nabors, M. W., and Stowe, B. B. (1967). Origin of indigo of woad. *Nature* (*London*) **216**, 547–549.

Esau, K. (1965). "Plant Anatomy," 2nd ed. Wiley, New York.

Esnault, R. (1968). Etude de l'action de l'acide β-indolyl-acétique sur le métabolisme de l'ARN de segments de coléoptiles d'avoine. *Bull. Soc. Chim. Biol.* **50**, 1887–1913.

Ettlinger, M. G., and Kjaer, A. (1968). Sulfur compounds in plants. *Recent Advan. Phytochem.* **1**, 59–144.

Evans, H. J., and Scott, D. (1964). Influence of DNA synthesis on the production of chromatid aberrations by X-rays and maleic hydrazide in *Vicia faba*. *Genetics* **49**, 17–38.

Evans, L. T., ed. (1969a). "The Induction of Flowering, Some Case Histories." Cornell Univ. Press, Ithaca, New York.

Evans, L. T. (1969b). A short history of the physiology of flowering. *In* "The Induction of Flowering, Some Case Histories" (L. T. Evans, ed.), pp. 1–13. Cornell Univ. Press, Ithaca, New York.

Evans, M. L., and Ray, P. M. (1969). Timing of the auxin response in coleoptiles and its implications regarding auxin action. *J. Gen. Physiol.* **53**, 1–20.

Evenari, M. (1949). Germination inhibitors. *Bot. Rev.* **15**, 153–194.

Evenari, M. (1961). Chemical influences of other plants. *In* "Handbuch der Pflanzenphysiologie" (W. Ruhland, ed.), Vol. XVI, pp. 691–736. Springer, Berlin.

Fambrough, D. M., Fujimura, F., and Bonner, J. (1968). Quantitative distribution of histone components in pea plant. *Biochemistry* **7**, 575–585.

Fawcett, C. H., and Spencer, D. M. (1969). Natural antifungal compounds. *In* "Fungicides" (D. C. Torgeson, ed.), Vol. 2, pp. 637–669. Academic Press, New York.

Fawcett, C. H., Taylor, H. F., Wain, F. L., and Wightman, F. (1956). The degradation of certain phenoxy acids, amides, and nitriles within plant tissues. *In* "The Chemistry and

Mode of Action of Plant Growth Substances" (R. L. Wain and F. Wightman, eds.), pp. 187–194. Butterworths, London and Washington, D. C.

Feierabend, J. (1969). Der Einfluss von Cytokininen auf die Bildung von Photosyntheseenzymen in Roggenkeimlingen. *Planta* **84**, 11–29.

Filner, P., and Varner, J. E. (1967). A test for de novo synthesis of enzymes: Density labeling with H_2O^{18} of barley α-amylase induced by gibberellic acid. *Proc. Nat. Acad. Sci. U.S.* **58**, 1520–1526.

Filner, P., Wray, J. L., and Varner, J. E. (1969). Enzyme induction in higher plants. *Science* **165**, 358–367.

Fishbein, L., Flamm, W. G., and Falk, H. L., eds. (1970). "Chemical Mutagens." Academic Press, New York.

Fitting, H. (1909). Beeinflussung der Orchideenblüten durch die Bestäubung und durch andere Umstände. *Z. Bot.* **1**, 1–86.

Fitting, H. (1910). Weitere entwicklungsphysiologische Untersuchungen an Orchideenblüten. *Z. Bot.* **2**, 225–267.

Fowden, L., Lewis, D., and Tristram, H. (1967). Toxic amino acids: Their action as antimetabolites. *Advan. Enzymol.* **29**, 89–163.

Fowden, L., Smith, I. K., and Dunhill, P. M. (1968). Some observations on the specificity of amino acid biosynthesis and incorporation into plant proteins. *In* "Recent Aspects of Nitrogen Metabolism in Plants" (E. J. Hewitt and C. V. Cutting, eds.), pp. 165–177. Academic Press, New York.

Fox, J. E. (1966). Incorporation of a kinin, N^6-benzyladenine into soluble RNA. *Plant Physiol.* **41**, 75–82.

Fox, J. E. (1969). The cytokinins. *In* "The Physiology of Plant Growth and Development" (M. B. Wilkins, ed.), pp. 85–123. McGraw-Hill, New York.

Fox, J. E., and Chen, C. -M. (1967). Characterization of labelled ribonucleic acid from tissue grown on C^{14} containing cytokinins. *J. Biol. Chem.* **242**, 4490–4494.

Fox, J. E. (1970). Active forms of the cytokinins. *Seventh International Conference on Plant Growth Substances* No. 23, *December 7–12* (Abstr.). *Canberra, Australia.*

Fox, J. J., Watanabe, K. A., and Bloch, A. (1966). Nucleoside antibiotics. *Progr. Nucl. Acid Res. Mol. Biol.* **5**, 251–313.

Fragata, M. (1970). The mitotic apparatus. A possible site of action of gibberellic acid. *Naturwissenschaften* **57**, 139–140.

Frear, D. E. H., ed. (1969). "Pesticide Handbook-Entoma." College Sci. Publ., State College, Pennsylvania.

Frederick, S. E., Newcomb, E. H., Vigil, E. L., and Wergin, W. P. (1968). Fine-structural characterization of plant microbodies. *Planta* **81**, 229–252.

Freed, V. H. (1966). Chemistry of herbicides. *In* "Pesticides and Their Effects on Soils and Water" (M. E. Bloodworth, ed.), pp. 25–43. Soil Sci. Soc. Amer., Madison, Wisconsin.

Fregda, A., and Åberg, B. (1965). Stereoisomerism in plant growth regulators of the auxin type. *Annu. Rev. Plant Physiol.* **16**, 53–72.

Fuller, W., and Hodgson, A. (1967). Conformation of the anticodon loop in tRNA. *Nature* (*London*) **215**, 817–821.

Furuya, M. (1968). Biochemistry and physiology of phytochrome. *Progr. Phytochem.* **1**, 347–405.

Gall, J. (1963). Chromosomes and cytodifferentiation. *In* "Cytodifferentiation and Macromolecular Synthesis" (M. Locke, ed.), pp. 119–143. Academic Press, New York.

Galston, A. W. (1967a). Changing the environment. Herbicides in Vietnam. II. *Scientist Citizen* **9**, 122–129.

Galston, A. W. (1967b). Regulatory systems in higher plants. *Amer. Sci.* **55**, 144–160.

Galston, A. W. (1968). Microspectrophotometric evidence for phytochrome in plant nuclei. *Proc. Nat. Acad. Sci. U.S.* **61**, 454–460.

Galston, A. W. (1970). Plants, people and politics. *BioScience* **20**, 405–410.

Galston, A. W., and Davies, P. J. (1969). Hormonal regulation in higher plants. *Science* **163**, 1288–1297.

Galston, A. W., and Davies, P. J. (1970). "Control Mechanisms in Plant Development." Prentice-Hall, Englewood Cliffs, New Jersey.

Galston, A. W., and Purves, W. K. (1960). The mechanism of action of auxin. *Annu. Rev. Plant. Physiol.* **11**, 239–276.

Galston, A. W., Lavee, S., and Siegel, B. Z. (1968). The induction and repression of peroxidase isozymes by 3-indoleacetic acid. *In* "Biochemistry and Physiology of Plant Growth Substances" (F. Wightman and G. Setterfield, eds.), pp. 455–472. Runge Press, Ottawa.

Gane, R. (1934). Production of ethylene by some ripening fruits. *Nature* (*London*) **134**, 1008.

Garb, S. (1961). Differential growth-inhibitors produced by plants. *Bot. Rev.* **27**, 422–443.

Garner, W. W., and Allard, H. A. (1920). Effect of the relative length of day and night and other factors of the environment on growth and reproduction. *J. Agr. Res.* **18**, 553–606.

Gaskin, P., and MacMillan, J. (1968). Plant hormones. VII. Identification and estimation of abscisic acid in a crude plant extract by combined gas chromatography-mass spectrometry. *Phytochemistry* **7**, 1699–1701.

Gautheret, R. J. (1942). "Manuel technique de culture des tissus végétaux; son état actuel, comparison avec la culture des tissus animaux." Hermann, Paris.

Gautheret, R. J. (1955). Sur la variabilité des propriétés physiologiques des cultures de tissus végétaux. *Année Biol.* [3] **31**, 145–171.

Gautheret, R. J. (1959). "La culture des tissus végétaux." Masson, Paris.

Gayler, K. R., and Glasziou, K. T. (1969). Plant enzyme synthesis: Hormonal regulation of invertase and peroxidase synthesis in sugar cane. *Planta* **84**, 185–194.

Geissman, T. A., Deuel, P., Bonde, E. K., and Addicott, F. A. (1954). Xanthinin: A plant growth regulatory compound from *Xanthium pennsylvanicum* I. *J. Amer. Chem. Soc.* **76**, 685–687.

Gerdemann, J. W. (1968). Vesicular-arbuscular mycorrhiza and plant growth. *Annu. Rev. Phytopathol.* **6**, 397–418.

Gibbs, M. (1970). The inhibition of photosynthesis by oxygen. *Amer. Sci.* **58**, 634–640.

Giertych, M. M. (1964). Endogenous growth regulators in trees. *Bot. Rev.* **30**, 292–311.

Gifford, E. M., Jr., and Nitsch, J. P. (1969). Responses of tobacco pith nuclei to growth substances. *Planta* **85**, 1–10.

Gilbert, W., and Müller-Hill, B. (1966). Isolation of the lac repressor. *Proc. Nat. Acad. Sci. U.S.* **56**, 1891–1898.

Goddard, D. R. (1945). The nature of respiration. *In* "Physical Chemistry of Cells and Tissues" (R. Höber, *et al.*, eds.), pp. 371–444. Blakiston, Philadelphia, Pennsylvania.

Goebel, K. (1908). "Einleitung in die Experimentelle Morphologie der Pflanzen." Teubner, Leipzig.

Goeschl, J. D., and Pratt, H. K. (1968). Regulatory roles of ethylene in the etiolated growth habit of *Pisum sativum*. *In* "Biochemistry and Physiology of Plant Growth Substances" (F. Wrightman and G. Setterfield, eds.), pp. 1299–1242. Runge Press, Ottawa.

Goethe, J. W. (1952). "Botanical Writings" (transl. by Bertha Muller, with an introduction by C. J. Engard). Univ. of Hawaii Press, Honolulu.

Goldsmith, M. H. M. (1968). The transport of auxin. *Annu. Rev. Plant Physiol.* **19**, 347–360.

Goldsmith, M. H. M. (1969). Transport of plant growth regulators. *In* "The Physiology of Plant Growth and Development" (M. B. Wilkins, ed.), pp. 127–162. McGraw-Hill, New York.

Graham, D., Grieve, A. M., and Smillie, R. M. (1968). Phytochrome as the primary photoregulator of the synthesis of Calvin cycle enzymes in etiolated pea seedlings. *Nature* (*London*) **218**, 89–90.

Graham, D., Hatch, M. D., Slack, C. K., and Smillie, R. M. (1970). Light-induced formation of enzymes of the C_4-dicarboxylic acid pathway of photosynthesis in detached leaves. *Phytochemistry* **9,** 521–532.

Granick, S., and Gibor, A. (1967). The DNA of chloroplasts, mitochondria and centrioles. *Progr. Nucl. Acid Res. Mol. Biol.* **6**, 143–186.

Green, P. B. (1969). Cell morphogenesis. *Annu. Rev. Plant Physiol.* **20**, 365–394.

Gregory, L. E. (1965). Physiology of tuberization in plants. *In* "Handbuch der Pflanzenphysiologie" (W. Ruhland, ed.), Vol. XV, Part 1, pp. 1928–1954. Springer, Berlin.

Greulach, V. A., and Plyler, D. B. (1966). Influence of 22 maleic hydrazide derivatives or related compounds on the growth of *Phaseolus vulgaris*. *J. Elisha Mitchell Scientific Society* **82**, 18–25.

Grümmer, G. (1961). The role of toxic substances in the interrelationship between higher plants. *Symp. Soc. Exp. Biol.* **15**, 209–228.

Grunwald, C., Mendez, J., and Stowe, B. B. (1968). Substrates for the optimum gas chromatographic separation of indolic methyl esters and the resolution of components of methyl 3-indolepyruvate solution. *In* "Biochemistry and Physiology of Plant Growth Substances" (F. Wightman and G. Setterfield, eds.), pp. 163–171. Runge Press, Ottawa.

Guern, J. (1970). Regulation of the cytokinin level inside the cell by enzymes of the adenosine deaminase type. *Seventh International Conference on Plant Growth Substances* No. 33, *December 7–12* (Abstr.). *Canberra, Australia.*

Guha, S., and Maheshwari, S. C. (1966). Cell division and differentiation of embryos in the pollen grains of *Datura* in vitro. *Nature* (*London*) **212**, 97–98.

Gurdon, J. B. (1968). Nucleic acid synthesis in embryos and its bearing on cell differentiation. *In* "Essays in Biochemistry" (P. N. Campbell and G. D. Greville, eds.), Vol. 4, pp. 25–68. Academic Press, New York.

Gurdon, J. B. (1969). Intracellular communication in early animal development. *Develop. Biol. Suppl.* **3**, 59–82.

Haagen-Smit, A. J., Dandliker, W. B., Wittwer, S. H., and Murneek, A. E. (1946). Isolation of 3-indoleacetic acid from immature corn kernels. *Amer. J. Bot.* **33**, 118–119.

Haber, A. H. (1968). Ionizing radiations as research tools. *Annu. Rev. Plant Physiol.* **19**, 463–489.

Haber, A. H., and Foard, D. E. (1964). Interpretations concerning cell division and growth. *In* "Régulateurs naturels de la croissance végétale" (J. P. Nitsch, ed.), pp. 491–503. C. N. R. S., Paris.

Haberlandt, G. (1913). Zur Physiologie der Zellteilung. *Sitzungsber, Deut. Akad. Wiss. Berlin, Kl. Math, Phys.* No. I, pp. 318–345.

Haberlandt, G. (1921). Wundhormone als Erreger von Zellteilung. *Beitr. Allg. Bot.* **2**, 1–53.

Hall. R. H. (1970). N^6-(Δ^2-isopentenyl)adenosine: Chemical reactions, biosynthesis, metabolism, and significance to the structure and function of tRNA. *Progr. Nucl. Acid Res. Mol. Biol.* **10**, 57–86.

Hall, R. H., and Srivastava, B. I. S. ((1968). Cytokinin activity of compounds obtained from soluble RNA. *Life Sci.* **7**, 7–13.

Hall, R. H., Chen, C-M., McLennan, B. D., and Werstiuk, E. (1970). N^6-(Δ^2-isopentenyl)-adenosine: Biosynthesis and metabolism in cytokinin-requiring tobacco pith tissue. p. 83 (Abstr.). *Eleventh Intern. Botan. Congr. Seattle, Washington.*

Hallaway, M., and Osborne, D. J. (1969). Ethylene: A factor in defoliation induced by auxins. *Science* **163**, 1067–1068.

Halperin, W. (1969). Morphogenesis in cell cultures. *Annu. Rev. Plant Physiol.* **20**, 395–418.

Hamner, K. (1963). Endogenous rhythms in controlled environments. *In* "Environmental Control of Plant Growth" (L. T. Evans, ed.), pp. 215–232. Academic Press, New York.

Hansen, E. (1966). Post harvest physiology of fruits. *Annu. Rev. Plant Physiol.* **17**, 459–480.

Hanson, K. R., Zucker, M., and Sondheimer, E. (1967). The regulation of phenolic biosynthesis and the metabolic roles of phenolic compounds in plants. *In* "Phenolic Compounds Metabolic Regulation" (B. J. Finkle and V. C. Runeckles, eds.), pp. 68–93. Appleton, New York.

Hardy, R. W. F., and Knight, E., Jr. (1968). Biochemistry and postulated mechanisms of nitrogen fixation. *Progr. Phytochem.* **1**, 407–489.

Harris, H. (1968). "Nucleus and Cytoplasm." Oxford Univ. Press (Clarendon), London and New York.

Hartsema, A. M. (1961). Influence of temperatures on flower formation and flowering of bulbous and tuberous plants. *In* "Handbuch der Pflanzenphysiologie" (W. Ruhland, ed.), Vol. XVI, pp. 123–167. Springer, Berlin.

Hashimoto, S., Miyazki, M., and Takemura, S. (1969). Nucleotide sequence of tyrosine transfer RNA from *Torulopsis utilis*. *J. Biochem.* (*Tokyo*) **65**, 659–661.

Haupt, W. (1968). Die Orientierung der Phytochrom Molekule in der *Mougeotia*-Zelle. *Z. Pflanzenphysiol.* **58**, 331–346.

Hause, L. L., Pattillo, R. A., Sances, Jr., A., and Mattingly, R. F. (1970). Cell surface coatings and membrane potentials of malignant and nonmalignant cells. *Science* **169**, 601–603.

Hecht, S. M., Leonard, N. J., Burrows, W. J., Armstrong, D. J., Skoog, F., and Occolowitz, J., (1969a). Cytokinin of wheat germ transfer RNA: 6-(4-hydroxy-3-methyl-2-butenylamino)-2-methylthio-9-β-D-ribofuranosylpurine. *Science* **166**, 1272–1274.

Hecht, S. M., Leonard, N. J., Occolowitz, J., Burrows, W. J., Armstrong, D. J., Skoog, F., Bock, R. M., Gillam, I., and Tener, G. M. (1969b). Cytokinins: Isolation and identification of 6-(3-methyl-2-butenylamino)-9-β-D-ribofuranosylpurine (2iPA) from yeast cysteine tRNA. *Biochem. Biophys. Res. Commun.* **35**, 205–209.

Hecht, S. M., Gupta, A. S., and Leonard, N. J. (1969c). Mass spectra of nucleoside components. *Anal. Biochem.* **30**, 249–270.

Hecht, S. M., Leonard, N. J., Schmitz, R. Y., and Skoog, F. (1970a). Cytokinins: Synthesis and growth promoting activity of 2-substituted compounds in the N^6-isopentenyladenine and zeatin series. *Phytochemistry* **9**, 1173–1180.

Hecht, S. M., Leonard, N. J., Schmitz, R. Y., and Skoog, F. (1970b). Cytokinins; Influence of side-chain planarity of N^6-substituted adenines and adenosines on their activity in promoting cell growth. *Phytochemistry* **9**, 1907–1913.

Heftmann, E. (1970). Insect molting hormones in plants. *Recent Adv. Phytochem.* **3**, 211–227.

Helgeson, J. P. (1968). The cytokinins. *Science* **161**, 974–981.

Helgeson, J. P., Krueger, S. M., and Upper, C. D. (1969). Control of logarithmic growth rates of tobacco callus tissue by cytokinins. *Plant Physiol.* **44**, 193–198.

Hendricks, S. B. (1959). The photoreaction and associated changes of plant photomorphogenesis. *In* "Photoperiodism and Related Phenomena in Plants and Animals," Publ. No. 55, pp. 423–438. Am. Assoc. Advanc. Sci., Washington, D. C.

Hendricks, S. B. (1969). Light in plant and animal development. *Develop. Biol. Suppl.* **3**, 227–243.

Hendricks, S. B., and Borthwick, H. A. (1963). Control of plant growth by light. *In* "Environmental Control of Plant Growth" (L. T. Evans, ed.), pp. 233–263. Academic Press, New York.

Hendricks, S. B., and Borthwick, H. A. (1967). The function of phytochrome in regulation of plant growth. *Proc. Nat. Acad. Sci. U.S.* **58**, 2125–2130.

Hendricks, S. B., and Siegelman, H. W. (1967). Phytochrome and photoperiodism in plants. *In* "Comprehensive Biochemistry" (M. Florkin and E. H. Stotz, eds.), Vol. 27, pp. 211–235. Elsevier, Amsterdam.

Hertel, R., Evans, M. L., Leopold, A. C., and Sell, H. M. (1969). The specificity of the auxin transport system. *Planta* **85**, 238–249.

Heslop-Harrison, J. (1956). The experimental modification of sex expression in flowering plants. *Biol. Rev.* **32**, 38–90.

Heslop-Harrison, J. (1960). Suppressive effects of 2-thiouracil on differentiation and flowering in *Cannabis sativa*. *Science* **132**, 1943–1944.

Heslop-Harrison, J. (1964). The control of flower differentiation and sex expression. *In* "Regulateurs naturels de la croissance végétale" (J. P. Nitsch, ed.), pp. 649–664. C. N. R. S, Paris.

Heslop-Harrison, J. (1967). Differentiation. *Annu. Rev. Plant Physiol.* **18**, 325–348.

Hillman, W. S. (1967). The physiology of phytochrome. *Annu. Rev. Plant Physiol.* **18**, 301–324.

Hillman, W. S. (1969). Photoperiodism and vernalization. *In* "The Physiology of Plant Growth and Development" (M. B. Wilkins, ed.), pp. 559–601. McGraw-Hill, New York.

Hindawi, I. J. (1970). "Air Pollution Injury to Vegetation." U. S. Dept. of Health, Education, and Welfare. Washington, D. C.

Hinman, R. L., and Lang, J. (1965). Peroxidase-catalyzed oxidation of indole-3-acetic acid. *Biochemistry* **4**, 144–158.

Hitchcock, A. E., and Zimmerman, P. W. (1937/8). The use of green tissue test objects for determining the physiological activity of growth substances. *Contrib. Boyce Thompson Inst.* **9**, 463–516.

Hoffman-Ostenhof, O. (1969). Enzymes involved in the *O*-methylation of inositols in higher plants, yeasts, and bacteria. *Ann. N. Y. Acad Sci.* **165**, 624–629.

Holley, R. W., Apgar, J., Everett, G. A., Madison, J. T., Marquisee, M., Merrill, S. H., Penswick, J. R., and Zamir, A. (1965). Structure of a ribonucleic acid. *Science* **147**, 1462–1465.

Holsten, R. D., Sugii, M., and Steward, F. C. (1965). Direct and indirect effects of radiation on plant cells: Their relation to growth and growth induction. *Nature* (*London*) **208**, 850–856.

Howe, K. J., and Steward, F. C. (1962). Anatomy and development of *Mentha piperita* L. In growth, nutrition and metabolism of *Mentha piperita* L. *Cornell Univ., Agr. Exp. Sta., Mem.* **379**, 11–40.

Huffman, C. W., Godar, E. M., Ohki, K., and Torgeson, D. C. (1968). Synthesis of hydrazine derivatives as plant growth inhibitors. *Agr. Food Chem.* **16**, 1041–1046.

Hutchinson, A., Taper, C. D., and Towers, G. H. N. (1959). Studies of phloridzin in *Malus*. *Can. J. Biochem. Physiol.* **37**, 901–910.

Huxley, J. S. (1935). Chemical regulation and the hormone concept. *Biol. Rev.* **10**, 427–441.

Hwang, Y.-S., and Matsui, M. (1968). Synthesis of the stereoisometric mixture of the compound having the proposed structure for "Auxin b lactone." *Agr. Biol. Chem.* **32**, 81–87.

Ilan, J., and Quastel, J. H. (1966). Effects of colchicine in nucleic acid metabolism during metamorphosis of *Tenebrio molitor* L., and in some mammalian tissues. *Biochem. J.* **100**, 448–457.

Ingle, J., and Key, J. L. (1965). A comparative evaluation of the synthesis of DNA-like RNA in excised and intact plant tissues. *Plant Physiol.* **40**, 1212–1219.

Ingle, J., Key, J. L., and Holm, R. E. (1965). Demonstration and characterization of a DNA-like RNA in excised plant tissue. *J. Mol. Biol.* **11**, 730–746.

Iriuchijima, S., and Tamura, S. (1970). Stereochemistry of pyrethrosin, cyclopyrethrosin acetate and isocyclopyrethrosin acetate. *Agr. Biol. Chem.* **34**, 204–209.

Israel, H. W., and Steward, F. C. (1966). The fine structure of quiescent and growing carrot cells: Its relation to growth induction. *Ann. Bot.* (*London*) [N. S.] **30**, 63–79.

Israel, H. W., and Steward, F. C. (1967). The fine structure and development of plastids in cultured cells of Daucus carota. *Ann. Bot.* (*London*) [N. S.] **31**, 1–18.

Israel, H. W., Salpeter, M. M., and Steward, F. C. (1968). The incorporation of radioactive proline into cultured cells: Interpretations based on radioautography and electron microscopy. *J. Cell Biol.* **39**, 698–715.

Israel, H. W., Mapes, M. O., and Steward, F. C. (1969). Pigments and plastids of totipotent carrot cells. *Amer. J. Bot.* **56**, 910–917.

Itai, C., Richmond, A., and Vaadia, Y. (1968). The role of root cytokinins during water and salinity stress. *Isr. J. Bot.* **17**, 187–195.

IUPAC-IUB Combined Commission on Biochemical Nomenclature. (1970). Abbreviations and symbols for nucleic acids, polynucleotides and their constituents. *Eur. J. Biochem.* **15,** 203–208.

Ivens, G. W., and Blackman, G. E. (1949). The effects of phenylcarbamates on the growth of higher plants. *Symp. Soc. Exp. Biol.* **3,** 266–282.

Izawa, S., and Good, N. E. (1965). The number of sites sensitive to 3-(3,4-dichlorophenyl)-1,1-dimethylurea, 3-(4-chlorophenyl)-1,1-trimethylurea and 2-chloro-4-(2-propylamino)-6-ethylamino-S-triazine in isolated chloroplasts. *Biochim. Biophys. Acta* **102**, 20–38.

Jachymczyk, W. J., and Cherry, J. H. (1968). Studies on messenger RNA from peanut plants: *In vitro* polyribosome formation and protein synthesis. *Biochim. Biophys. Acta* **157**, 368–377.

Jackson, M. B., and Osborne, D. J. (1970). Ethylene, the natural regulation of leaf abscission. *Nature* (*London*) **225**, 1019–1022.

Jacob, F., and Monod, J. (1963). Genetic repression, allosteric inhibition, and cellular differentiation. *In* "Cytodifferentiation and Macromolecular Synthesis" (M. Locke, ed.), pp. 30–64. Academic Press, New York.

Jacobs, W. P. (1970). Regeneration and differentiation of sieve tube elements. *Int. Rev. Cytol.* **28**, 239–273.

Jacobson, J. S., and Hill, A. C., eds. (1970). "Recognition of Air Pollution Injury: A Pictorial Atlas." Air Pollution Control Agency, Raleigh, North Carolina.

Jaffe, L. F. (1969). On the centripetal course of development, the *Fucus* egg, and self-electrophoresis. *Develop. Biol. Suppl.* **3**, 83–111.

James, C. S., and Wain, R. L. (1968). Studies on plant growth-regulating substances. *Ann. Appl. Biol.* **61**, 295–302.

Johansen, D. A. (1950). "Plant Embryology." Chronica Botanica, Waltham, Massachusetts.

John, B., and Lewis, K. R. (1965). The meiotic system. *Protoplasmatologia* **6 (F1)**, 1–335.

John, B., and Lewis, K. R. (1968). The chromosome complement. *Protoplasmatologia* **6(A)**, 1–206.

John, B., and Lewis, K. R. (1969). The chromosome cycle. *Protoplasmatologia* **6(B)**, 1–125.

Johri, M. M., and Varner, J. E. (1967). Gibberellins. *In* "Methods in Developmental Biology" (F. H. Wilt and N. K. Wessells, eds.), pp. 595–611. Crowell, New York.

Johri, M. M., and Varner, J. E. (1968). Enhancement of RNA synthesis in isolated pea nuclei by gibberellic acid. *Proc. Nat. Acad. Sci. U.S.* **59**, 269–276.

Jones, B. L., and Varner, J. (1967). The bioassay of gibberellins. *Planta* **72**, 155–161.

Kearney, P. C., and Kaufman, D. D., eds. (1969). "Degradation of Herbicides." Marcel Dekker, New York.

Keeler, R. F. (1969). Toxic and teratogenic alkaloids of western range plants. *Agr. Food Chem.* **17**, 473–482.

Kefeli, V. L., Kof, E. M., Knipl, Ya. S., Bukhanova, L. V., and Yarviste, E. L. (1969). Conversion of isosalipurpurposide and phloridzin on contact with various plant tissues. *Biochemistry* **34**, 719–726 (translated from the Russian, *Biokhimiya*).

Kefford, N. P., Bruce, M. I., and Zwar, J. A. (1966). Cytokinin activities of phenylurea derivatives—Bud growth. *Planta* **68**, 292–296.

Kefford, N. P., Zwar, J. A., and Bruce, M. I. (1968). Antagonism of purine and urea cytokinin activities by derivatives of benzylurea. *In* "Biochemistry and Physiology of Plant Growth Substances" (F. Wightman and G. Setterfield, eds.), pp. 61–69. Runge Press, Ottawa.

Kende, H., and Tavares, J. E. (1968). On the significance of cytokinin incorporation into RNA. *Plant Physiol.* **43**, 1244–1248.

Kepes, A. (1967). Sequential transcription and translation in the lactose operon of *Escherichia coli*. *Biochim. Biophys. Acta* **138**, 107–123.

Kersten, H., and Kersten, W. (1969). Inhibitors acting on DNA and their use to study DNA replication and repair. *In* "Inhibitors, Tools in Cell Research" (T. Bücher and H. Sies, eds.), pp. 11–31. Springer, New York.

Key, J. L. (1969). Hormones and nucleic acid metabolism. *Annu. Rev. Plant Physiol.* **20**, 449–474.

Key, J. L., and Ingle, J. (1968). RNA metabolism in response to auxin. *In* "Biochemistry and Physiology of Plant Growth Substances" (F. Wightman and G. Setterfield, eds.), pp. 711–722. Runge Press, Ottawa.

Khan, A. A. (1967). Physiology of morphactins: Effect on gravi- and photo-response. *Physiol. Plant.* **20**, 306–313.

Kihlman, B. A. (1966). "Actions of Chemicals on Dividing Cells." Prentice-Hall, Englewood Cliffs, New Jersey.

King, L. J. (1966). "Weeds of the World." Leonard Hill, London.

Kjaer, A. (1960). Naturally derived isothiocyanates (mustard oils) and their parent glucosides. *Fortschr. Chem. Org. Naturst.* **18**, 122–176.

Klämbt, D., and Kovoor, A. (1969). Cytokinins and transfer ribonucleic acids. I. Presence of cytokinins in various amino-acid specific transfer ribonucleic acids of baker's yeast. *Physiol. Plant.* **22**, 453–457.

Klebs, G. (1918). Uber die Blütenbildung von *Sempervivum*. *Flora (Jena)* **111**, 128–151.

Kline, L. K., Fittler, F., and Hall, R. H. (1969). N^6-(Δ^2-isopentenyl)adenosine. Biosynthesis in transfer ribonucleic acid *in vitro*. *Biochemistry* **8**, 4361–4371.

Koblitz, H. (1969a). Enzymologische Untersuchungen an pflanzlichen Gewebekulturen. I. Zur Frage der Beeinflussung im Fermentsystemem durch Gibberellin und Chlorochlinchlorid-Eine Übersicht. *Biol. Zentralbl.* **88**, 283–294.

Koblitz, H. (1969b). Enzymologische Untersuchungen an pflanzlichen Gewebekulturen. II. Die Wirkung von Gibberellinsaüre und Chlorocholinchlorid auf die Aktivität einiger Fermentsysteme *in vitro* kultivierter Pflanzengewebe. *Biol. Zentralbl.* **88**, 409–423.

Köhler, D. (1969). Phytochromabhängiger Ionentransport in Erbsensämlingen. *Planta* **84**, 158–165.

Korn, E. D. (1969). Cell membranes: Structure and synthesis. *Annu. Rev. Biochem.* **38**, 263–288.

Koshimizu, K., Fukui, H., Inui, M., Ogawa, Y., and Mitsui, T. (1968a). Gibberellin A 23 in immature seeds of *Lupinus luteus*. *Tetrahedron Lett.* **9**, 1143–1147.

Koshimizu, K., Inui, M., Fukui, H., and Mitsui, T. (1968b). Isolation of (+)-abscisyl-β-D-glucopyranoside from immature fruit of *Lupinus luteus*. *Agr. Biol. Chem.* **32**, 789–791.

Koshimizu, K., Kobayashi, A., Fujita, T., and Mitsui, T. (1968c). Structure-activity relationships in optically active cytokinins. *Phytochemistry* **7**, 1989–1994.

Krelle, E., and Libbert, E. (1967). Wirkung eines Morphaktins auf die Amylase-synthese in Gerstenendosperm. *Planta* **76**, 179–181.

Krieg, D. R. (1963). Specificity of chemical mutagenesis. *Progr. Nucl. Acid Res. Mol. Biol.* **2**, 125–168.

Krikorian, A. D., and Berquam, D. L. (1969). Plant cell and tissue cultures: The role of Haberlandt. *Bot. Rev.* **35**, 58–88.

Krikorian, A. D., and Steward, F. C. (1969). Biochemical differentiation: The biosynthetic potentialities of growing and quiescent tissue. *In* "Plant Physiology: A Treatise" (F. C. Steward, ed.), Vol. 5B, pp. 227–326. Academic Press, New York.

Kroeger, H. (1967). Hormones, ion balances and gene activity. *Mem. Soc. Endocrinol.* **15**, 55–66.

Kuhn, R., and Löw, I. (1954). Die Konstitution des Solanins. *Angew. Chem.* **66**, 639–640.

Kupchan, S. M. (1970). Recent advances in the chemistry of tumor inhibitors of plant origin. *Trans. N. Y. Acad. Sci.* [2] **32**, 85–106.

Kupchan, S. M., Aynehchi, Y., Cassidy, J. M., McPhail, A. T., Sim, G. A., Schnoes, H. K., and Burlingame, A. L. (1966). The isolation and structural elucidation of two novel sesquiterpenoid tumor inhibitors from *Elephantopus elatus* Bertol. *J. Amer. Chem. Soc.* **88**, 3674–3676.

Kurosawa, E. (1926). Experimental studies on the secretion of "bakanae" fungus on rice plants. *Trans. Natur. Hist. Soc. Formosa* **16**, 213–227; translated *in* "Source Book on Gibberellin, 1828–1957" (F. H. Stodola, ed.), pp. 111–128. Agr. Res. Serv., U. S. Dept. of Agr., Peoria, Illinois, 1958.

Kutáček, M. (1967). Indolederivate in Pflanzen der Familie Brassicaceae. *Wiss. Z. Univ. Rostock, Math.-Naturwiss. Reihe* **16**, 417–426.

Lamport, D. T. A. (1965). The protein component of primary cell walls. *Advan. Bot. Res.* **2**, 151–218.

Lamport, D. T. A. (1970). Cell wall metabolism. *Annu. Rev. Plant Physiol.* **21**, 235–270.

Lang, A. (1957). The effect of gibberellin upon flower formation. *Proc. Nat. Acad. Sci. U.S.* **43**, 709–717.

Lang, A. (1965). Physiology of flower initiation. *In* "Handbuch der Pflanzenphysiologie" (W. Ruhland, ed.), Vol. XV, Part 1, pp. 1380–1536. Springer, Berlin.

Lang, A. (1966). Intercellular regulation in plants. *In* "Major Problems in Developmental Biology" (M. Locke, ed.), pp. 251–287. Academic Press, New York.

Lang, A. (1970). Gibberellins: Structure and metabolism. *Annu. Rev. Plant Physiol.* **21**, 537–570.

Laycock, D. G., and Hunt, J. A. (1969). Synthesis of rabbit globin by a bacterial cell free system. *Nature* (*London*) **221**, 1118–1122.

Leonard, N. J., Hecht, S. M., Skoog, F., and Schmitz, R. Y. (1968). Cytokinins: Synthesis and biological activity of related derivatives of 2iP, 3iP, and their ribosides. *Isr. J. Chem.* **6**, 539–550.

Leonard, N. J., Hecht, S. M., Skoog, F., and Schmitz, R. (1969). Cytokinins: Synthesis, mass spectra, and biological activity of compounds related to zeatin. *Proc. Nat. Acad. Sci. U.S.* **63**, 175–182.

Lerman, L. S. (1961). Structural considerations in the interaction of DNA and acridenes. *J. Mol. Biol.* **3**, 18–30.

Leshem, Y. (1970). Gonadotropin-like hormone as an endogenous plant growth regulator. *Seventh International Conference on Plant Growth Substances* No. 46, *December 7–12* (Abstr.). *Canberra, Australia.*

Leshem, Y., Avtalion, R. R., Schwartz, M., and Kahana, S. (1969). Presence and possible mode of action of a proteinaceous gonadotropin-like growth regulating factor in plant systems. *Plant Physiol.* **44**, 75–77.

Letham, D. S. (1966). Regulators of cell division in plant tissues. II. A cytokinin in plant extracts: Isolation and interaction with other growth regulators. *Phytochemistry* **5**, 269–286.

Letham, D. S. (1967a). Chemistry and physiology of kinetin-like compounds. *Annu. Rev. Plant Physiol.* **18**, 349–364.

Letham, D. S. (1967b). Regulators of cell division in plant tissues. V. A comparison of the activities of zeatin and other cytokinins in five bioassays. *Planta* **74**, 228–242.

Letham, D. S. (1968). A new cytokinin bioassay and the naturally occurring cytokinin complex. *In* "Biochemistry and Physiology of Plant Growth Substances" (F. Wightman and G. Setterfield, eds.), pp. 19–30. Runge Press, Ottawa.

Letham, D. S., Mitchell, R. E., Cebalo, T., and Stanton, D. W. (1969). Regulators of cell division in plant tissues. VII. The synthesis of zeatin and related 6-substituted purines. *Aust. J. Chem.* **22**, 205–219.

Lipetz, J. (1970). Wound healing in higher plants. *Int. Rev. Cytol.* **27**, 1–28.

Lipmann, F. (1969). Polypeptide chain elongation in protein biosynthesis. *Science* **164**, 1024–1031.

Loening, V. E. (1968). RNA structure and metabolism. *Annu. Rev. Plant Physiol.* **19**, 37–70.

Loewus, F. (1969). Metabolism of inositol in higher plants. *Ann. N. Y. Acad. Sci.* **165**, 577–598.

Lonberg-Holm, K. K. (1967). Nucleic acid synthesis in seedlings. *Nature (London)* **213**, 454–457.

Luckwill, L. C. (1968). Relations between plant growth regulators and nitrogen metabolism. *In* "Recent Aspects of Nitrogen Metabolism in Plants" (E. J. Hewitt and C. V. Cutting, eds.), pp. 189–199. Academic Press, New York.

Lüttze, V., Bauer, K., and Köhler, D. (1968). Frühwirkungen von gibberellinsäure auf Membrantransporte in Jungen Erbsenpflanzen. *Biochim. Biophys. Acta* **150**, 452–459.

Luyten, I., Joustra, G., and Blaauw, A. H. (1926). The results of temperature-treatment in summer for the Darwin tulip. II. *Proc., Kon. Ned. Akad. Wetensch.* **29**, 113–126.

McClintock, B. (1951). Chromosome organization and genic expression. *Cold Spring Harbor Symp. Quant. Biol.* **16**, 13–47.

McClintock, B. (1967). Genetic systems regulating gene expression during development. *Develop. Biol. Suppl.* **1**, 84–112.

McCready, C. C. (1966). Translocation of growth regulators. *Annu. Rev. Plant Physiol.* **17**, 283–294.

MacDougal, D. T. (1926). Growth and permeability of century old cells. *Amer. Natur.* **60**, 393–415.

MacDougal, D. T., and Long, F. L. (1927). Characteristics of cells attaining great age. *Amer. Natur.* **61**, 385–406.

MacDougal, D. T., and Smith, G. M. (1927). Long-lived cells of the redwood. *Science* **66**, 456–457.

MacMillan, J., and Pryce, R. J. (1968). Phaseic acid, a putative relative of abscisic acid, from seed of *Phaseolus multiflorus*. *Chem. Commun.* No. 3, 124–126.

MacMillan, J., and Takahashi, N. (1968). Proposed procedure for the allocation of trivial names to gibberellins. *Nature (London)* **217**, 170–171.

Maheshwari, P. (1950). "An Introduction to the Embryology of Angiosperms." McGraw-Hill, New York.

Mann, J. D., Yung, K. -H., Storey, W. B., Pu, M., and Conley, J. (1967). Similarity between phytokinins and herbicidal urethanes. *Plant Cell Physiol.* **8**, 613–622.

Mapson, L. W. (1969). Biogenesis of ethylene. *Biol. Rev.* **44**, 155–187.

Martello, O. J., Woo, S. L. C., Reimann, E. M., and Davis, E. W. (1970). Effect of protein kinase on ribonucleic acid polymerase. *Biochemistry* **9**, 4807–4813.

Marumo, S., Abe, H., Hattori, H., and Munakata, K. (1968). Isolation of a novel auxin, methyl 4-chloroindoleacetate from immature seeds of *Pisum sativum*. *Agr. Biol. Chem.* **32**, 117–118.

Masuda, Y. (1968). Role of cell-wall-degrading enzymes in cell-wall loosening in oat coleoptiles. *Planta* **83**, 171–184.

Mathan, D. S. (1965). Phenylboric acid, a chemical agent simulating the effect of the lanceolate gene in the tomato. *Amer. J. Bot.* **52**, 185–192.

Matsubara, S., Armstrong, D. J., and Skoog, F. (1968). Cytokinins in tRNA of *Corynebacterium fascians*. *Plant Physiol.* **43**, 451–453.

Mayer, A. M., and Poljakoff-Mayber, A. (1963). "The Germination of Seeds." Pergamon Press, Oxford.

Mayer, J. (1967). Starvation as a weapon. Herbicides in Vietnam. I. *Scientist Citizen* **9**, 115–121.

Mayo, F. R., and Walling, C. (1940). The peroxide effect in the addition of reagents to unsaturated compounds and in rearrangement ractions. *Chem. Rev.* **27**, 351–412.

"Mechanism of Protein Synthesis." (1969). *Cold Spring Harbor Symp. Quant. Biol.* **34**.

Melchers, G., and Lang, A. (1948). Die Physiologie der Blütenbildung. *Biol. Zentralbl.* **67**, 105–174.

Menzie, C. M. (1966). Metabolism of pesticides. *U.S., Fish Wildl. Serv., Spec. Sci. Rep.: Wildl.* **96**.

Milborrow, B. V. (1970). The metabolism of abscisic acid. *J. Exp. Bot.* **21**, 17–29.

Miller, C. O. (1963). Kinetin and kinetin-like compounds. *In* "Moderne Methoden der Pflanzenanalyze" (K. Paech and M. V. Tracey, eds.), Vol. 6, pp. 194–202. Springer, Berlin.

Miller, C. O. (1967). Cytokinins. *In* "Methods in Developmental Biology" (F. H. Wilt and N. K. Wessells, eds.), pp. 613–622. Crowell, New York.

Miller, C. O. (1968). Naturally-occurring cytokinins. *In* "Biochemistry and Physiology of Plant Growth Substances" (F. Wightman and G. Setterfield, eds.), pp. 33–45. Runge Press, Ottawa.

Miller, C. O. (1969). Control of deoxyisoflavone synthesis in soybean tissue. *Planta* **87**, 26–35.

Miller, C. O. (1970). Plant hormones. *In* "Biochemical Actions of Hormones" (G. Litwack, ed.), Vol. 1, pp. 503–518, Academic Press, New York.

Miller, C. O., Skoog, F., von Saltza, M. H., and Strong, F. M. (1955a). Kinetin, a cell division factor from deoxynucleic acid. *J. Amer. Chem. Soc.* **77**, 1392.

Miller, C. O., Skoog, F., Okumura, F. S., von Saltza, M. H., and Strong, F. M. (1955b). Structure and synthesis of kinetin. *J. Amer. Chem. Soc.* **77**, 2662–2663.

Mitchell, J. W., and Livingston, G. A. (1968). Methods of studying plant hormones and growth-regulating substances. *U.S., Dep. Agr., Agr. Handb.* **336**.

Miura, G. A., and Miller, C. O. (1969). Cytokinins from a variant strain of cultured soybean cells. *Plant Physiol.* **44**, 1035–1039.

Mohan Ram, H. Y., Ram, M., and Steward, F. C. (1962). Growth and development of the banana plant. III. A. The origin of the inflorescence and the development of the flowers. B. The structure and development of the fruit. *Ann. Bot.* (*London*) [N. S.] **26**, 657–673.

Mohr, G., and Ziegler, H., eds. (1969). "Symposium über Morphaktine." Fischer, Stuttgart.

Mohr, H. (1966). Differential gene activation as a mode of action of phytochrome 730. *Photochem. Photobiol.* **5**, 469–483.

Mohr, H. (1969). Photomorphogenesis. *In* "The Physiology of Plant Growth and Development" (M. B. Wilkins, ed.), pp. 509–556. McGraw-Hill, New York.

Monod, J. (1966). From enzymatic adaptation to allosteric transitions. *Science* **154**, 475–483.

Monod, J., Wyman, J., and Changeux, J. -P. (1965). On the nature of allosteric transitions: A plausible model. *J. Mol. Biol.* **12**, 88–118.

Moreland, D. E. (1967). Mechanisms of action of herbicides. *Annu. Rev. Plant Physiol.* **18**, 365–386.

Moreland, D. E., Egley, G. H., Worsham, A. D., and Monaco, T. J. (1966). Regulation of plant growth by constituents from higher plants. *Advan. Chem. Ser.* **53**, 112–141.

Morimoto, H., Sanno, Y., and Oshio, H. (1966). Chemical studies on heliangine, a new sesquiterpene lactone isolated from the leaves of *Helianthus tuberosus*. *Tetrahedron* **22**, 3173–3179.

Morré, D. J., and Key, J. L. (1967). Auxins. *In* "Methods in Developmental Biology" (F. H. Wilt and N. K. Wessells, eds.), pp. 575–593. Crowell, New York.

Mosbach, K. (1969). Biosynthesis of lichen substances, products of a symbiotic association. *Angew. Chem. Int. Edit.* **8**, 240–250.

Mothes, K. (1964). The role of kinetin in plant regulation. *In* "Régulateurs naturels de la croissance végétale" (J. P. Nitsch, ed.), pp. 131–140. C. N. R. S., Paris.

Mousseron-Canet, M., Mani, J. -C., Durand, B., Nitsch, J., Dornand, J., and Bonnafous, J. -C. (1970). Analogues de l'acide abscisique (±) hormone de dormance. Relations structure-activité. *C. R. Acad. Sci., Ser. D* **270**, 1936–1939.

Muller, C. H. (1966). The role of chemical inhibition (allelopathy) in vegetational composition. *Bull. Torrey Bot. Club* **93**, 332–351.

Muller, C. H. (1970). Phytotoxins as plant habitat variables. *Recent Adv. Phytochem.* **3**, 105–121.

Murayama, A., and Tamura, S. (1970a). Uber Fragin, ein Neues Biologisch Aktives Stoffwechselproduct von *Pseudomonas fragi*. II. Zur Struktur und Chemie des Fragins. *Agr. Biol. Chem.* **34**, 122–129.

Murayama, A., and Tamura, S. (1970b). Fragin, a new biologically active metabolite of a *Pseudomonas*. Part III. Synthesis of (±)-Fragin. *Agr. Biol. Chem.* **34**, 130–134.

Murayama, A., Hata, K., and Tamura, S. (1969). Fragin, a new biologically active metabolite of *Pseudomonas*. Part I. Isolation, characterization and biological activities. *Agr. Biol. Chem.* **33**, 1599–1605.

Murneek, A. E. (1948). History of research in photoperiodism. *In* "Vernalization and Photoperiodism" (A. E. Murneek and R. O. Whyte, eds.), pp. 39–61. Chronica Botanica, Waltham, Massachusetts.

Nagl, W. (1969). Banded polytene chromosomes in the legume *Phaseolus vulgaris*. *Nature (London)* **221**, 70–71.

Nakamura, T., Takahashi, N., Matsui, M., and Hwang, Y. -S. (1966). Activity of synthesized auxin b lactone as the plant growth regulator of auxin type. *Plant Cell Physiol.* **7**, 693–696.

Nakata, K., and Tanaka, M. (1968). Differentiation of embryoids from developing germ cells in anther culture of tobacco. *Jap. J. Genet.* **43**, 67–71.

Needham, J. (1931). "Chemical Embryology." Cambridge Univ. Press, London and New York.

Neely, P. M., and Phinney, B. O. (1957). The use of the mutant dwarf-1 of maize as a quantitative bioassay for gibberellin activity. *Plant Physiol.* **32**, Suppl., xxxi.

Nelson, C. D. (1963). Effect of climate on the distribution and translocation of assimilates. *In* "Environmental Control of Plant Growth" (L. T. Evans, ed.), pp. 149–174. Academic Press, New York.

Nelson, G. (1970). "Environmental Warfare in Vietnam," Congressional Record—Senate, 116, No. 22, pp. 1982–1984.

Nicholls, P. B. (1967). The isolation of indole-3-acetyl-*O*-myoinositol from *Zea mays*. *Planta* **72**, 258–264.

Nicholls, P. G., and Paleg, L. G. (1963). A barley endosperm bioassay for gibberellins. *Nature (London)* **199**, 823–824.

Nichols, J. L. (1970). Nucleotide sequence from the polypeptide chain termination region of the coat protein cistron in bacteriophage R_{1c}RNA. *Nature (London)* **225**, 147–151.

Nickell, L. G. (1955). Effects of antigrowth substances in normal and atypical plant growth. *In* "Antimetabolites and Cancer," pp. 129–151. Am. Assoc. Advance. Sci., Washington, D. C.

Niizeki, H., and Oono, K. (1968). Induction of haploid rice plant from another culture. *Proc. Jap. Acad.* **44**, 554–557.

Nikolaeva, M. G. (1969). "Physiology of Deep Dormancy in Seeds" (transl. from Russian). Israel Program Sci. Transl., Jerusalem.

Nishikawa, M., Kamiya, K., Takabatake, A., and Oshio, H. (1966). The X-ray analysis of dihydroheliangine monochloroacetate. *Tetrahedron* **22**, 3601–3606.

Nitsch, J. P. (1950). Growth and morphogenesis of the strawberry as related to auxin. *Amer. J. Bot.* **37**, 211–215.

Nitsch, J. P. (1963). The mediation of climatic effects through endogenous regulating substances. *In* "Environmental Control of Plant Growth" (L. T. Evans, ed.), pp. 175–193. Academic Press, New York.

Nitsch, J. P. (1968). Studies on the mode of action of auxins, cytokinins and gibberellins at the subcellular level. *In* "Biochemistry and Physiology of Plant Growth Substances"

(F. Wightman and G. Setterfield, eds.), pp. 563–580. Runge Press, Ottawa.

Nitsch, J. P., and Nitsch, C. (1962). Composés phénoliques et croissance végétale. *Ann. Physiol. Veg.* **4**, 211–225.

Nitsch, J. P., and Nitsch, C. (1969). Haploid plants from pollen grains. *Science* **163**, 85–87.

Noodén, L. D., and Thimann, K. V. (1966). Action of inhibitors of RNA and protein synthesis on cell enlargement. *Plant Physiol.* **41**, 157–164.

Northcote, D. H. (1969). The synthesis and metabolic control of polysaccharides and lignin during the differentiation of plant cells. *In* "Essays in Biochemistry" (P. N. Campbell and G. D. Greville, eds.), Vol. 5, pp. 90–137. Academic Press, New York.

Nougarède, A. (1967). Experimental cytology of the shoot apical cells during vegetative growth and flowering. *Int. Rev. Cytol.* **21**, 203–351.

O'Brien, T. J., Jarvis, B. C., Cherry, J. H., and Hanson, J. B. (1968). Enhancement by 2,4-dichlorophenoxyacetic acid of chromatin RNA polymerase in soybean hypocotyl tissue. *Biochim. Biophys. Acta* **169**, 35–43.

Okamoto, T., and Torii, Y. (1968). Lycoricidinol and lycoricidine, new plant-growth regulators in bulbs of *Lycoris radiata* Herb. *Chem. Pharm. Bull.* **16**, 1860–1864.

Okamoto, T., Isogai, Y., Koizumi, T., Nishino, T., and Satoh, Y. (1970). Studies on plant-growth regulators. IV. Structure of growth retardants and syntheses of related compounds. *Chem. Pharm. Bull.* **18**, 243–248.

Orians, G. H., and Pfeiffer, E. W. (1970). Ecological effects of the war in Vietnam. *Science* **168**, 544–554.

Oritani, T., and Yamashita, K. (1970). Studies on abscisic acid. Part II. The oxidation products of methyl α- and β-cyclocitrylidene-acetates and methyl α-cyclogeranate. *Agr. Biol. Chem.* **34**, 198–203.

Osborne, D. J. (1968). Defoliation and defoliants. *Nature (London)* **219**, 564–567.

Owens, L. D. (1969). Toxins in plant disease: Structure and mode of action. *Science* **165**, 18–25.

Padilla, G. M., Cameron, I. L., and Whitson, G. L. (1969). "The Cell Cycle." Academic Press, New York.

Paleg, L. G. (1965). Physiological effects of gibberellins. *Annu. Rev. Plant Physiol.* **16**, 291–322.

Paleg, L. G., Kende, H., Ninnemann, H., and Lang, A. (1965). Physiological effects of gibberellic acid. VII. Growth retardants on barley endosperm. *Plant Physiol.* **40**, 165–169.

Palmer, R. L., Lewis, L. N., Hield, H. Z., and Kumamoto, J. (1967). Abscission induced by Betahydroxyethylhydrazine: Conversion of Betahydroxyethylhydrazine to ethylene. *Nature (London)* **216**, 1216–1217.

Pardee, A. B. (1968). Membrane transport proteins. *Science* **162**, 632–637.

Partanen, C. R. (1959). Quantitative chromosomal changes and differentiation in plants. *In* "Developmental Cytology" (D. Rudnick, ed.), pp. 21–45. Ronald Press, New York.

"Pesticides and Their Effects on Soils and Water." (1966). ASA Spec. Publ. No. 8. Soil Sci. Soc. Am. Madison, Wisconsin.

Peterkofsky, A., and Jesensky, C. (1969). The localization of N^6-(Δ^2-isopentenyl)adenosine among the acceptor species of transfer ribonucleic acid of *Lactobacillus acidophilus*. *Biochemistry* **8**, 3798–3809.

Peters, R. A. (1954). Biochemical light upon an ancient poison: A lethal synthesis. *Endeavour* **13**, 147–154.

Peters, R. A., Hall, R. J., Ward, P. F. V., and Sheppard, N. (1960). The chemical nature of the toxic compounds containing fluorine in the seeds of *Dichapetalum toxicarium*. *Biochem. J.* **77**, 17–23.

Peterson, G. E. (1967). The discovery and development of 2,4-D. *Agr. Hist. Ser.* **41**, 243–253.

Phinney, B. O. (1956). Growth response of single-gene dwarf mutants in maize to gibberellic acid. *Proc. Nat. Acad. Sci. U.S.* **42**, 185–189.

Phinney, B. O. (1969). Biosynthesis and metabolism of gibberellin. *In* "Gibberellin, Chemistry, Biochemistry and Physiology" (S. Tamura, ed.), pp. 195–219. Tokyo Univ. Press, Tokyo (in Japanese).

Phinney, B. O., West, C. A., Ritzel, M., and Neely, P. M. (1957). Evidence for "gibberellin-like" substances from flowering plants. *Proc. Nat. Acad. Sci. U.S.* **43**, 398–404.

Picard, C. (1968). "Aspects et mécanismes de la vernalisation." Masson, Paris.

Pilet, P. E. (1961). "Les phytohormones de croissance; méthodes, chimie, biochemie, physiologie, applications practique." Masson, Paris.

Pilet, P. E., and Gaspar, T. (1968). "Le catabolisme auxinique." Masson, Paris.

"Plant Growth Regulators." (1968). Compromising papers (with discussions) read at a joint symposium organized by the Pesticides Group of the S. C. I. and the Phytochemical Society. Soc. Chem. Ind., London.

Pogo, B. G. T., Allfrey, V. G., and Mirsky, A. E. (1966). RNA synthesis and histone acetylation during the course of gene activation in lymphocytes. *Proc. Nat. Acad. Sci. U. S.* **55**, 805–812.

Pollard, J. K., and Steward, F. C. (1959). The use of C^{14}-proline by growing cells: Its conversion to protein and to hydroxyproline. *J. Exp. Bot.* **10**, 17–32.

Pollard, J. K., Shantz, E. M., and Steward, F. C. (1961). Hexitols in coconut milk: Their role in the nurture of dividing cells. *Plant Physiol.* **36**, 492–501.

Popham, R. A. (1958). Some causes underlying cellular differentiation. *Ohio J. Sci.* **56**, 347–353.

Porter, W. L., and Thimann, K. V. (1965). Molecular requirements for auxin action. I. Halogenated indoles and indoleacetic acid. *Phytochemistry* **4**, 229–243.

Pratt, H. K., and Goeschl, J. D. (1969). Physiological roles of ethylene in plants. *Annu. Rev. Plant Physiol.* **20**, 541–584.

Prescott, D. M. (1964). *In* "Synchrony in Cell Division and Growth" (E. Zeuthen, ed.), pp. 71–100. Wiley (Interscience), New York.

Raggio, M., and Raggio, N. (1962). Root nodules. *Annu. Rev. Plant Physiol.* **13**, 109–128.

Rappaport, L., and Wolf, N. (1969). The problem of dormancy in potato tubers and related structures. *Symp. Soc. Exp. Biol.* **23**, 219–240.

Rasmussen, H. (1970). Cell communication, Calcium ion, and cyclic adenosine monophosphate. *Science* **170**, 404–412.

Ray, P. M. (1958). Destruction of auxin. *Annu. Rev. Plant Physiol.* **9**, 81–118.

Ray, P. M. (1969). The action of auxin in cell enlargement in plants. *Develop. Biol. Suppl.* **3**, 172–205.

Rayle, D. L., and Cleland, R. (1970). Enhancement of wall loosening and elongation by acid solutions. *Plant Physiol.* **46**, 250–253.

Rayle, D. L., Evans, M. L., and Hertel, R. (1970a). Action of auxin on cell elongation. *Proc. Nat. Acad. Sci. U. S.* **65**, 184–191.

Rayle, D. L., Haughton, P. M., and Cleland, R. (1970b). An *in vitro* system that simulates plant cell extension growth. *Proc. Nat. Acad. Sci. U. S.* **67**, 1814–1817.

Redemann, C. T., Rappaport, L., and Thompson, R. H. (1968). Phaseolic acid: A new plant growth regulator from bean seeds. *In* "Biochemistry and Physiology of Plant Growth Substances" (F. Wightman and G. Setterfield, eds.), pp. 109–124. Runge Press, Ottawa.

Reich, E., and Goldberg, I. H. (1964). Actinomycin and nucleic acid function. *Progr. Nucl. Acid Res. Mol. Biol.* **3**, 184–230.

Reid, D. M., and Clements, J. B. (1968). RNA and protein synthesis; prerequisites of red light-induced gibberellin synthesis. *Nature* (*London*) **219**, 607–609.

Reid, D. M., Clements, J. B., and Carr, D. J. (1968). Red light induction of gibberellin synthesis in leaves. *Nature* (*London*) **217**, 580–582.

Reinert, J. (1968). Morphogenese in Gewebe-und Zellkulturen. *Naturwissenschaften* **55**, 170–175.

"Replication of DNA in Micro-Organisms" (1968). *Cold Spring Harbor Symp. Quant.* ***Biol.*** **33**,

Report of the Secretary. (1969). "Commission on Pesticides and their Relationships to Environmental Health," Parts I and II, Chapter 8, pp. 657–675. U. S. Dept. of Health, Education, and Welfare, Washington, D. C.

Report to the Subcommittee on Science, Research, and Development of the Committee on Science and Astronautics. U. S. House of Representatives, 91st Congress. (1969). "A Technology Assessment of the Vietnam Defoliant Matter." U. S. Govt. Printing Office, Washington, D. C.

Rice, E. L. (1965). Inhibition of nitrogen-fixing and nitrifying bacteria by seed plants. *Physiol. Plant.* **18**, 255–268.

Rice, E. L. (1967). Chemical warfare between plants. *Bios* **38**, 67–74.

Rich, P., Waggoner, P. E., and Tomlinson, H. (1970). Ozone uptake by bean leaves. *Science* **169**, 79–80.

Richards, F. J. (1951). Phyllotaxis: Its quantitative expression and relation to growth in the apex. *Phil. Trans. Roy. Soc. London, Ser. B* **235**, 509–564.

Richards, F. J., and Schwabe, W. W. (1969). Phyllotaxis: A problem of growth and form. *In* "Plant Physiology: A Treatise" (F. C. Steward, ed.), Vol. 5A, pp. 79–116. Academic Press, New York.

Rier, J. P., and Beslow, D. T. (1967). Sucrose concentration and the differentiation of xylem in callus. *Bot. Gaz.* **128**, 73–77.

Roberts, E. H. (1969). Seed dormancy and oxidation processes. *Symp. Soc. Exp. Bol.* **23**, 161–192.

Robison, G. A., Butcher, R. W., and Sutherland, E. W. (1968). Cyclic AMP, *Annu. Rev. Biochem.* **37**, 149–174.

Rogozińska, J. H. (1967). Triacanthine, growth substances and the *in vitro* culture of *Gleditsia triacanthos* L. *Bull. Acad. Pol. Sci., Ser. Sci. Biol.* **15**, 313–317.

Rogozińska, J. H., Helgeson, J. P., and Skoog, F. (1964). Tests for kinetin-like growth promoting activities of triacanthine and its isomers, 6-(γ-γ-dimethyllallylamino)-purine. *Physiol. Plant.* **17**, 165–176.

Rothfield, L., and Finkelstein, A. (1968). Membrane biochemistry. *Annu. Rev. Biochem.* **37**, 463–496.

Rothwell, K., and Wright, S. T. C. (1967). Phytokinin activity in some new 6-substituted purines. *Proc. Roy. Soc., Ser. B* **167**, 202–223.

Rubinstein, B., Drury, K. S., and Park, R. B. (1969). Evidence for bound phytochrome in oat seedlings. *Plant Physiol.* **44**, 105–109.

Rüdiger, W. (1969). Uber die Struktur des Phytochrom-Chromophors und seine Protein-Bindung. *Justus Liebigs Ann. Chem.* **723**, 208–212.

Sachs, J. (1887). "Lectures on the Physiology of Plants" (transl. by H. M. Ward). Oxford Univ. Press (Clarendon), London and New York.

Sadgopal, A. (1968). The genetic code after the excitement. *Advan. Genet.* **14**, 325–404.

Sam, J., and Valentine, J. L. (1969). Preparation and properties of 2-benzoxazolinones. *J. Pharm. Sci.* **58**, 1043–1054.

Sarkissian, I. V. (1968). Nature of molecular action of 3-indoleacetic acid. *In* "Biochemistry and Physiology of Plant Growth Substances" (F. Wightman and G. Setterfield, eds.), pp. 473–485. Runge Press, Ottawa.

Satomura, Y., and Sato, A. (1965). Isolation and physiological activity of sclerin, a metabolite of *Sclerotinia* fungus. *Agr. Biol. Chem.* **29**, 337–344.

Schneider, G. (1970). Morphactins: Physiology and performance. *Annu. Rev. Plant Physiol.* **21**, 499–536.

Schüepp, O. (1966). "Meristeme; Wachstum und Formbildung in den Teilungsgeweben höherer Pflanzen." Birkhäuser, Basel.

Schuetz, R. D., and Titus, R. L. (1967). Benzo[b]thiophene derivatives. I. 6-methoxybenzo[b]-thiophene analogs of plant growth regulators. *J. Heterocycl. Chem.* **4**, 465–468.

Sequeira, L., Hemingway, R. J., and Kupchan, S. M. (1968). Vernolepin: A new, reversible plant growth inhibitor. *Science* **161**, 789–790.

Shannon, J. S., and Letham, D. S. (1966). Regulators of cell division in plant tissues. IV. The mass spectra of cytokinins and other 6-aminopurines. *N. Z. J. Sci.* **9**, 833–842.

Shantz, E. M. (1966). Chemistry of naturally-occurring growth-regulating substances. *Annu. Rev. Plant Physiol.* **17**, 409–438.

Shantz, E. M., and Steward, F. C. (1955). The identification of Compound A from coconut milk as 1,3-diphenylurea. *J. Amer. Chem. Soc.* **77**, 6351–6353.

Shantz, E. M., and Steward, F. C. (1964). The growth-stimulating complexes of coconut milk, corn, and *Aesculus*. *In* "Regulateurs naturels de la croissance végétale" (J. P. Nitsch, ed.), pp. 59–75. C. N. R. S., Paris.

Shantz, E. M., and Steward, F. C. (1968). A growth substance from the vesicular embryo sac of *Aesculus*. *In* "Biochemistry and Physiology of Plant Growth Substances" (F. Wightman and G. Setterfield, eds.), pp. 893–909. Runge Press, Ottawa.

Shantz, E. M., Steward, F. C., Smith, M. S., and Wain, R. L. (1955). Investigations on growth and metabolism. VI. Growth of potato tuber tissue in culture: The synergistic action of coconut milk and synthetic growth regulating compounds. *Ann. Bot.* (*London*) [N. S.] **19**, 49–58.

Shaw, G., Smallwood, B. M., and Steward, F. C. (1968). Synthesis and cytokinin activity of the 3-, 7- and 9-methyl derivatives of zeatin. *Experientia* **24**, 1089.

Shaw, G., Smallwood, B. M., and Steward, F. C. (1971). The structure and physiological activity of some N^6-substituted adenines and related compounds. *Phytochemistry* (in press).

Shelanski, M. L., and Taylor, E. W. (1967). Isolation of a protein subunit from microtubules. *J. Cell Biol.* **34**, 549–554.

Shih, T. Y., and Bonner, J. (1970). Template properties of DNA-polypeptide complexes. *J. Mol. Biol.* **50**, 333–344.

Shimokawa, K., and Kasai, Z. (1968). A possible incorporation of ethylene into RNA in Japanese morning glory seedlings. *Agr. Biol. Chem.* **32**, 680–682.

Siegelman, H. W., and Hendricks, S. B. (1964). Phytochrome and its control of plant growth and development. *Advan. Enzymol.* **26**, 1–33.

Siegelman, H. W., Chapman, D. J., and Cole, W. J. (1968). The bile pigments of plants. *Biochem. Soc. Symp.* **28**, 107–120.

Singer, B., and Fraenkel-Conrat, H. (1969). The role of conformation in chemical mutagenesis. *Progr. Nucl. Acid. Res. Mol. Biol.* **9**, 1–29.

Singleton, V. L., and Kratzer, F. H. (1969). Toxicity and related physiological activity of phenolic substances of plant origin. *Agr. Food Chem.* **17**, 497–511.

Skoog, F., and Armstrong, D. J. (1970). Cytokinins. *Annu. Rev. Plant Physiol.* **21**, 359–384.

Skoog, F., and Leonard, N. J. (1968). Sources and structure: Activity relationships of cytokinins. *In* "Biochemistry and Physiology of Plant Growth Substances" (F. Wightman and G. Setterfield, eds.), pp. 1–18. Runge Press, Ottawa.

Skoog, F., and Miller, F. O. (1957). Chemical regulation of growth and organ formation in plant tissues cultured *in vitro*. *Symp. Soc. Exp. Biol.* **11**, 118–131.

Skoog, F., Hamzi, H. Q., and Szweykowska, A. M. (1967). Cytokinins: Structure/activity relationships. *Phytochemistry* **6**, 1169–1192.

Smith, H., and Attridge, T. H. (1970). Increased phenylalanine ammonia-lyase activity due to light treatment and its significance for the mode of action of phytochrome. *Phytochemistry* **9**, 487–495.

Smith, R. H., and Murashige, T. (1970). *In vitro* development of the isolated shoot apical meristem of angiosperms. *Amer. J. Bot.* **57**, 562–568.

Smock, R. M. (1970). Environmental factors affecting ripening of fruit. *HortScience* **5**, 37–38.

Sondheimer, E. (1964). Chlorogenic acids and related depsides. *Bot. Rev.* **30**, 667–712.

Sondheimer, E., Michniewicz, B. M., and Powell, L. E. (1969). Biological and chemical properties of the epidioxide isomer of abscisic acid and its rearrangement products. *Plant Physiol.* **44**, 205–209.

Spencer, D., and Whitfield, P. R. (1967). Ribonucleic acid synthesizing activity of spinach chloroplasts and nuclei. *Arch. Biochem. Biophys.* **121**, 336–345.

Spencer, D. M. (1963). Antibiotics in seeds and seedling plants. *Proc. Easter Sch. Agr. Sci. Univ. Nottingham* **9**, 125–146.

Spencer, M. (1969). Ethylene in nature. *Fortschr. Chem. Org. Naturst.* **27**, 31–80.

Srivastava, B. I. S. (1967a). Effect of kinetin on biochemical changes in excised barley leaves and in tobacco pith tissue culture. *Ann. N. Y. Acad. Sci.* **144**, 260–278.

Srivastava, B. I. S. (1967b). Cytokinins in plants. *Int. Rev. Cytol.* **22**, 349–387.

Staal, G. B. (1967). Plants as sources of insect hormones. *Proc., Kon. Ned. Akad. Wetensch., Ser. C* **70**, 409–418.

Steffens, G. L., and Cathey, H. M. (1969). Selection of fatty acid derivatives: Surfactant Formulations for the control of plant meristems. *Agr. Food Chem.* **17**, 312–317.

Steffens, G. L., Tso, T. C., and Spaulding, D. W. (1967). Fatty alcohol inhibition of tobacco axillary and terminal bud growth. *Agr. Food Chem.* **15**, 972–975.

Steitz, J. A. (1969). Polypeptide chain initiation: Nucleotide sequences of the three ribosomal binding sites in bacteriophage R_{17} RNA. *Nature (London)* **224**, 957–964.

Stern, H. (1966). The regulation of cell division. *Annu. Rev. Plant Physiol.* **17**, 345–378.

Stern, H., and Hotta, Y. (1968). Biochemical studies of male gametogenesis in Liliaceous plants. *Curr. Top. Develop. Biol.* **3**, 37–63.

Steward, F. C. (1958). Growth and organized development of cultured cells. III. Interpretations of the growth from free cell to carrot plant. *Amer. J. Bot.* **45**, 709–713.

Steward, F. C. (1968). "Growth and Organization in Plants." Addison-Wesley, Reading, Massachusetts.

Steward, F. C. (1970). From cultured cells to whole plants: The induction and control of their growth and morphogenesis. *Proc. Roy. Soc., London Ser. B* **175**, 1–30.

Steward, F. C., and Bidwell, R. G. S. (1966). Storage pools and turnover systems in growing and non-growing cells: Experiments with C^{14}-sucrose, C^{14}-glutamine, and C^{14}-asparagine. *J. Exp. Bot.* **17**, 716–741.

Steward, F. C., and Caplin, S. M. (1952a). Investigations on growth and metabolism of plant cells. III. Evidence for growth inhibitors in certain mature tissues. *Ann. Bot. (London)* [N. S.] **16**, 477–489.

Steward, F. C., and Caplin, S. M. (1952b). Investigations on growth and metabolism of plant cells. IV. Evidence on the role of coconut milk factor in development. *Ann. Bot. (London)* [N. S.] **16**, 491–504.

Steward, F. C., and Degani, N. (1969). Endogenous characteristics of different clones of carrot explants and their exogenous requirements for growth. *Ann. Bot. (London)* [N. S.] **33**, 615–646.

Steward, F. C., and Mohan Ram, H. Y. (1961). Determining factors in cell growth: Some implications for morphogenesis in plants. *Advan. Morphog.* **1**, 189–265.

Steward, F. C., and Mott, R. L. (1970). Cell solutes and growth: Salt accumulation in plants re-examined. *Int. Rev. Cytol.* **28**, 275–370.

Steward, F. C., and Pollard, J. K. (1958). ^{14}C-proline and hydroxyproline in the protein metabolism of plants. *Nature* (*London*) **182**, 828–832.

Steward, F. C., and Rao, K. V. N. (1970). Investigations on the growth and metabolism of cultured explants of *Daucus carota*. III. The range of responses induced in carrot explants by exogenous growth factors and by trace elements. *Planta* **91**, 129–145.

Steward, F. C., and Shantz, E. M. (1956). The chemical induction of growth in plant tissue cultures. *In* "The Chemistry and Mode of Action of Plant Growth Substances" (R. L. Wain and F. Wightman, eds.), pp. 165–186. Butterworth, Academic Press, London and Washington, D. C.

Steward, F. C., and Shantz, E. M. (1959). The chemical regulation of growth (some substances and extracts which induce growth and morphogenesis). *Annu. Rev. Plant Physiol.* **10**, 379–404.

Steward, F. C., and Sutcliffe, J. F. (1959). Plants in relation to inorganic salts. *In* "Plant Physiology: A Treatise" (F. C. Steward, ed.), Vol. 2, pp. 253–478. Academic Press, New York.

Steward, F. C., Wright, R., and Berry, W. E. (1932). The absorption of an accumulation of solutes by living plant cells. III. The respiration of cut discs of potato tuber in air and immersed in water, with observations upon the surface: Volume effects and salt accumulation. *Protoplasma* **16**, 576–611.

Steward, F. C., Caplin, S. M., and Millar, F. K. (1952). Investigations on growth and metabolism of plant cells. I New techniques for the investigation of metabolism, nutrition and growth in undifferentiated cells. *Ann. Bot.* (*London*) [N. S.] **16**, 57–77.

Steward, F. C., Mapes, M. O., and Mears, K. (1958a). Growth and organized development of cultured cells. II. Organization in cultures grown from freely suspended cells. *Amer. J. Bot.* **45**, 705–708.

Steward, F. C., Mapes, M. O., and Smith, J. (1958b). Growth and organized development of cultured cells. I. Growth and development of freely suspended cells. *Amer. J. Bot.* **45**, 693–703.

Steward, F. C., Pollard, J. K., Patchett, A. A., and Witkop, B. (1958c). The effects of selected nitrogen compounds on the growth of plant tissue cultures. *Biochim. Biophys. Acta* **28**, 308–317.

Steward, F. C., Shantz, E. M., Mapes, M. O., Kent, A. E., and Holsten, R. D. (1964a). Growth-promoting substances from the environment of the embryo. I. The criteria and measurement of growth-promoting activity and the responses induced. *In* "Régulateurs naturels de la croissance végétale" (J. P. Nitsch, ed.), pp. 45–58. C. N. R. S., Paris.

Steward, F. C., Mapes, M. O., Kent, A. E., and Holsten, R. D. (1964b). Growth and development of cultured plant cells. *Science* **143**, 20–27.

Steward, F. C., Kent, A. E., and Mapes, M. O. (1966). The culture of free cells and its significance for embryology and morphogenesis. *Curr. Top. Develop. Biol.* **1**, 113–154.

Steward, F. C., Kent, A. E., and Mapes, M. O. (1967). Growth and organization in cultured cells: Sequential and synergistic effects of growth-regulating substances. *Ann. N. Y. Acad. Sci.* **144**, Art. 1, 326–334.

Steward, F. C., Israel, H. W., and Mapes, M. O. (1968a). Growth regulatory substances: Their roles observed at different levels of cellular organization. *In* "Biochemistry and Physiology of Plant Growth Substances" (F. Wightman and G. Setterfield, eds.), pp. 875–892. Runge Press, Ottawa.

Steward, F. C., Neumann, K. -H., and Rao, K. V. N. (1968b). Investigations on the growth and metabolism of cultured explants of *Daucus carota*. II. Effects of iron, molybdenum and manganese on metabolism. *Planta* **81**, 351–371.

Steward, F. C., Mapes, M. O., and Ammirato, P. V. (1969). Growth and morphogenesis in tissue and free cell cultures. *In* "Plant Physiology: A Treatise" (F. C. Steward, ed.), Vol. 5B, pp. 329–376. Academic Press, New York.

Steward, F. C., Mott, R. L., Israel, H. W., and Ludford, P. M. (1970). Proline in the vesicles and sporelings of *Valonia ventricosa* and the concept of cell wall protein. *Nature* (*London*) **225**, 760–762.

Steward, F. C., Ammirato, P. V., and Mapes, M. O. (1970). Growth and development of totipotent cells: Some problems, procedures and perspectives. *Ann. Bot.* (*London*) [N. S.] **34**, 761–787.

Stodola, F. H., ed. (1958). "Source Book on Gibberellin." Agr. Res. Serv., U. S. Dept. of Agriculture, Peoria, Illinois.

Stowe, B. (1959). Occurrence and metabolism of simple indoles in plants. *Fortschr. Chem. Org. Naturst.* **17**, 248–297.

Stowe, B., and Yamaki, T. (1957). The history and physiological action of the gibberellins. *Annu. Rev. Plant Physiol.* **8**, 181–216.

Stowe, B. B., and Hudson, V. W. (1969). Growth promotion in pea stem sections. III. By alkyl nitriles, alkyl acetylenes and insect juvenile hormones. *Plant Physiol.* **44**, 1051–1057.

Street, H. E. (1957). Excised root culture. *Biol. Rev.* **32**, 117–155.

Street, H. E. (1969). Growth in organized and unorganized systems. *In* "Plant Physiology: A Treatise" (F. C. Steward, ed.), Vol. 5B, pp. 3–224. Academic Press, New York.

Street, H. E., and Henshaw, G. G. (1966). Introduction and methods employed in plant tissue culture. *In* "Cells and Tissues in Culture" (E. N. Willmer, ed.), Vol. 3, pp. 459–532. Academic Press, New York.

Strong, F. M. (1958). Kinetin and kinins. "Topics in Microbial Chemistry," pp. 98–157. Wiley, New York.

Sunderland, N., and Brown, R. (1956). Distribution of growth in the apical region of the shoot of *Lupinus albus*. *J. Exp. Bot.* **7**, 127–145.

Sunderland, N., and Wicks, F. M. (1969). Cultivation of haploid plants from tobacco pollen. *Nature* (*London*) **224**, 1227–1229.

Sunderland, N., Heyes, J. K., and Brown, R. (1956). Growth and metabolism in the shoot apex of *Lupinus albus*. *In* "The Growth of Leaves" (F. L. Milthorpe, ed.), pp. 77–90. Butterworth, London and Washington, D. C.

Takahashi, N., and Curtis, R. W. (1961). Isolation and characterization of malformin. *Plant Physiol.* **36**, 30–36.

Tamura, S., and Nagao, M. (1969a). Synthesis of novel plant growth inhibitors structurally related to abscisic acid. *Agr. Biol. Chem.* **33**, 296–298.

Tamura, S., and Nagao, M. (1969b). 5-(1,2-epoxy-2,6,6-trimethyl-1-cyclohexyl)-3-methyl-*cis, trans*-2,4-pentadieneoic acid and its esters: New Plant growth inhibitors structurally related to abscisic acid. *Planta* **85**, 209–219.

Tamura, S., Suzuki, A., Yamauchi, M., Ogawa, Y., and Imamura, S. -i. (1967). Isolation of fumaric acid from silk worm pupas as a causative agent to release reducing sugars in embryoless rice endosperms. *Agr. Biol. Chem.* **31**, 1248–1250.

Tamura, S., Takahashi, N., Murofushi, N., Yokota, T., and Kato, J. (1968). Isolation of new gibberellins from higher plants and their biological activity. *In* "Biochemistry and Physiology of Plant Growth Substances" (F. Wightman and G. Setterfield, eds.), pp. 85–99. Runge Press, Ottawa.

Tanada, T. (1968). A rapid photoreversible response of barley root tips in the presence of 3-indoleacetic acid. *Proc. Nat. Acad. Sci. U. S.* **59**, 376–380.

Taniguchi, E., and White, G. A. (1967). Site of action of the phytotoxin, Helminthosporal. *Biochem. Biophys. Res. Commun.* **28**, 879–885.

Tanner, W. (1969). The function of *myo*-inositol glycosides in yeasts and higher plants. *Ann. N. Y. Acad. Sci.* **165**, 726–742.

Tata, J. R. (1968). Hormonal regulation of growth and protein synthesis. *Nature* (*London*) **219**, 331–337.

Tatum, E. L., Barratt, R. W., and Cutter, V. M. Jr. (1949). Chemical induction of colonial paramorphs in *Neurospora* and *Syncephalastrum*. *Science* **109**, 509–511.

Taylor, H. F. (1968). Carotenoids as possible precursors of abscisic acid in plants. *In* "Plant Growth Regulators," SCI Monogr. No. 31, pp. 22–35.

Taylor, H. F., and Burden, R. S. (1970a). Xanthoxin, a new naturally occurring plant growth inhibitor. *Nature* (London) **227**, 302–304.

Taylor, H. F., and Burden, R. S. (1970b). Identification of plant growth inhibitors produced by photolysis of violaxanthin. *Phytochemistry* **9**, 2217–2223.

Taylor, H. F., and Smith, T. A. (1967). Production of plant growth inhibitors from xanthophylls: A possible source of dormin. *Nature* (*London*) **215**, 1513–1514.

Temin, H. M., and Mizutani, S. (1970). RNA-dependent DNA polymerase in virions of Rous sarcoma virus. *Nature* (*London*) **226**, 1211–1213.

Tewari, K. K., and Wildman, S. G. (1968). Functional chloroplast DNA. I. Hybridization studies involving nuclear and chloroplast DNA with RNA from cytoplasmic (80S) and chloroplast (70S) ribosomes. *Proc. Nat. Acad. Sci. U.S.* **59**, 569–576.

Tezuka, T., and Yamamoto, Y. (1969). NAD kinase and phytochrome. *Bot. Mag.* **82**, 130–133.

Thimann, K. V. (1948). Plant growth hormones. *In* "The Hormones" (G. Pincus and K. V. Thimann, eds.), Vol. 1, pp. 5–119. Academic Press, New York.

Thimann, K. V. (1952). Plant growth hormones. *In* "The Action of Hormones in Plants and Invertebrates" (K. V. Thimann, ed.), pp. 1–76. Academic Press, New York.

Thimann, K. V. (1956). Promotion and inhibition: Twin themes of physiology. *Amer. Natur.* **90**, 145–162.

Thimann, K. V. (1965). Toward an endocrinology of higher plants. *Recent Progr. Horm. Res.* **21**, 579–596.

Thimann, K. V. (1967a). Phototropism. *In* "Comprehensive Biochemistry" (M. Florkin and E. H. Stotz, eds.), Vol. 27, pp. 1–29. Elsevier, Amsterdam.

Thimann, K. V. (1967b). Tropisms in plants. *Embryologia* **10,** 89–113.

Thimann, K. V. (1969). The auxins. *In* "The Physiology of Plant Growth and Development." (M. B. Wilkins, ed.), pp. 3–45. McGraw-Hill, New York.

Thimann, K. V., and Beth, K. (1959). Action of auxins on *Acetabularia* and the effect of enucleation. *Nature* (*London*) **183,** 946–948.

Thimann, K. V., and Laloraya, M. M. (1960). Changes in nitrogen in pea stem sections under the action of kinetin. *Physiol. Plant* **13**, 165–178.

Thimann, K. V., and Schneider, K. V. (1938). Differential growth in plant tissues. *Amer. J. Bot.* **25**, 627–641.

Thompson, H. C. (1953). Vernalization of growing plants. *In* "Growth and Differentiation in Plants" (W. E. Loomis, ed.), pp. 179–196. Iowa State Coll. Press, Ames, Iowa.

Tokoroyama, T., Maeda, S., Nishikawa, T., and Kubota, T. (1969). The synthesis of sclerin. *Tetrahedron* **25**, 1047–1054.

Tolbert, N. E. (1960). (2-chloroethyl)trimethylammonium chloride and related compounds as plant growth substances. *J. Biol. Chem.* **235**, 475–479.

Tomaszewski, M. (1964). The mechanism of synergistic effects between auxin and some natural phenolic substances. *In* "Régulateurs naturels de la croissance végétale" (J. P. Nitsch, ed.), pp. 335–351. C. N. R. S., Paris.

Toole, E. H., Hendricks, S. B., Borthwick, H. A., and Toole, V. K. (1956). Physiology of seed germination. *Annu. Rev. Plant Physiol.* **7**, 299–324.

Torrey, J. G. (1966). The initiation of organized development in plants. *Advan. Morphog.* **5**, 39–91.

Trewavas, A. J. (1968a). Relationship between plant growth hormones and nucleic acid metabolism. *Progr. Phytochem.* **1**, 113–160.

Trewavas, A. J. (1968b). Effect of IAA on RNA and protein synthesis. *Arch. Biochem. Biophys.* **123**, 324–335.

Tukey, H. B., ed. (1954). "Plant Regulators in Agriculture." Wiley, New York.

Tulecke, W. (1965). Haploidy versus diploidy in the reproduction of cell type. *In* "Reproduction: Molecular, Subcellular and Cellular" (M. Locke, ed.), pp. 217–241. Academic Press, New York.

Turetskaya, R., Kefeli, V., Kutácek, M., Vacková, K., Tschumakovski, N. G., and Krupnikova, T. (1968). Isolation and some physiological properties of natural plant growth inhibitors. *Biol. Plant.* **10**, 205–221.

Uribe, E. G. (1970). Phloretin: An inhibitor of phosphate transfer and electron flow in spinach chloroplasts. *Biochemistry* **9**, 2100–2106.

Valio, I. F. M., and Schwabe, W. W. (1970). Growth and dormancy in *Lunularia cruciata* L. *J. Exp. Bot.* **21**, 138–150.

van Overbeek, J. (1966). Plant hormones and regulators. *Science* **152**, 721–731.

van Overbeek, J., Conklin, M. E., and Blakeslee, A. F. (1941). Factors in coconut milk essential for growth and development of very young *Datura* embryos. *Science* **94**, 350–351.

van Overbeek, J., Loeffler, J. E., and Mason, M. I. (1967). Dormin (Abscisin II), Inhibitor of Plant DNA Synthesis? *Science* **156**, 1497–1499.

van Overbeek, J., Loeffler, J. E., and Mason, M. I. R. (1968). Mode of action of abscisic acid. *In* "Biochemistry and Physiology of Plant Growth Substances" (F. Wightman and G. Setterfield, eds.), pp. 1593–1607. Runge Press, Ottawa.

van Sumere, C., and Massart, L. (1958). Natural substances in relation to germination. *In* "Biochemistry of Antibiotics" (K. Y. Spitzy and R. Brunner, eds.), pp. 20–32. Pergamon Press, Oxford.

Van't Hof, J. (1968a). The action of IAA and kinetin on the mitotic cycle of proliferative and stationary phase excised root meristems. *Exp. Cell Res.* **51**, 167–176.

Van't Hof, J. (1968b). Experimental procedures for measuring cell population kinetic parameters in plant root meristems. *Methods Cell Physiol.* **3**, 95–117.

Vardar, Y., ed. (1968). "The Transport of Plant Hormones." North-Holland Publ., Amsterdam.

Varner, J. E., and Johri, M. M. (1968). Hormonal control of enzyme synthesis. *In* "Biochemistry and Physiology of Plant Growth Substances" (F. Wightman and G. Setterfield, eds.), pp. 793–814. Runge Press, Ottawa.

Vegis, A. (1964). Dormancy in higher plants. *Annu. Rev. Plant Physiol.* **15**, 185–224.

Veldstra, H. (1953). The relation of chemical structure to biological activity in growth substances. *Annu. Rev. Plant Physiol.* **4**, 151–198.

Villiers, T. A. (1968). An autoradiographic study of the effect of the plant hormone abscisic acid on nucleic acid and protein metabolism. *Planta* **82**, 342–354.

Virtanen, A. I., and Hietala, P. K. (1960). Precursors of benzoxazolinone in rye plants. I. Precursor II., The aglucone. *Acta Chem. Scand.* **14**, 499–502.

Vliegenthart, J. A., and Vliegenthart, J. F. G. (1966). Reinvestigation of authentic samples of auxins A and B and their related products by mass spectrometry. *Rec. Trav. Chim. Pays-Bas Belg.* **85**, 1266–1272.

Voeller, B. R. (1964). The plant cell: Aspects of its form and function. *In* "The Cell" (J. Brachet and A. E. Mirsky, eds.), Vol. 6, pp. 245–312. Academic Press, New York.

Waddington, C. H. (1957). "Strategy of the Genes." Macmillan, New York.

Wain, R. L., and Carter, A. (1967). Uptake, translocation and transformations by higher plants. *In* "Fungicides" (D. C. Torgeson, ed.), Vol. 1, pp. 561–611. Academic Press, New York.

Wain, R. L., and Fawcett, C. H. (1969). Chemical plant growth regulation. *In* "Plant Physiology: A Treatise" (F. C. Steward, ed.), Vol. 5A, pp.231–296. Academic Press, New York.

Wain, R. L., and Taylor, H. F. (1965). Phenols as plant growth regulators. *Nature* (*London*) **207**, 167–169.

Wallace, R. H. (1926). The production of intumescences upon apple twigs by ethylene gas. *Bull. Torrey Bot. Club* **53**, 385–402.

Waller, G. R., and Burström, H. (1969). Diterpenoid alkaloids as plant growth inhibitors. *Nature* (*London*) **222**, 576–578.

Wang, T. S. C., Yang, T. -K., and Chuang, T. -T. (1967). Some phenolic acids as plant growth inhibitors. *Soil Sci.* **103**, 239–246.

Wardlaw, C. W. (1953). Comparative observations on the shoot apices of vascular plants. *New Phytol.* **52**, 195–208.

Wardlaw, C. W. (1955). "Embryogenesis in Plants." Methuen, London.

Wardlaw, C. W. (1965a). Physiology of embryonic development in cormophytes. *In* "Handbuch der Pflanzenphysiologie" (W. Ruhland, ed.), Vol. XV, Part 1, pp. 844–965. Springer, Berlin.

Wardlaw, C. W. (1965b). The organization of the shoot apex. *In* "Handbuch der Pflanzenphysiologie" (W. Ruhland, ed.), Vol. XV, Part 1, pp. 966–1076. Springer, Berlin.

Wardlaw, C. W. (1965c). "Organization and Evolution in Plants." Longmans, Green, New York.

Wardlaw, C. W. (1968). "Morphogenesis in Plants." Methuen, London.

Wareing, P. F. (1969a). Germination and dormancy. *In* "The Physiology of Plant Growth and Development" (M. B. Wilkins, ed.), pp. 605–644. McGraw-Hill, New York.

Wareing, P. F. (1969b). The control of bud dormancy in seed plants. *Symp. Soc. Exp. Biol.* **23**, 241–262.

Wareing, P. F., and Ryback, G. (1970). Abscisic acid: A newly discovered growth-regulating substance in plants. *Endeavour* **29**, 84–88.

Waring, M. J. (1968). Drugs which affect the structure and function of DNA. *Nature* (*London*) **219**, 1320–1325.

Watson, J. D. (1970). "Molecular Biology of the Gene," 2nd edition. Benjamin, New York.

Weevers, T. (1949). Fifty years of plant physiology. "Scheltema & Holkema's Boekhandel en Uitgeversmaatschappij N. V.," Amsterdam.

Weidner, M., Rissland, I., and Mohr, H. (1968). Photoinduction of phenylalanine ammonia-lyase in mustard seedlings: Involvement of phytochrome. *Naturwissenschaften* **55**, 452.

Weidner, M., Rissland, I., Lohmann, L., and Mohr, H. (1969). Die Regulation der PAL-Aktivität durch Phytochrom in Senfkeimlingen (*Sinapis alba* L.). *Planta* **86**, 33–41.

Went, F. A. F. C. (1935). The investigations on growth and tropisms carried on in the Botanical Laboratory of the University of Utrecht during the last decade. *Biol. Rev.* **10**, 187–207.

Went, F. W. (1935). Auxin, the plant growth-hormone. *Bot. Rev.* **1**, 162–182.

Went, F. W., and Thimann, K. V. (1937). "The Phytohormones." Macmillan, New York.

West, C. A., and Upper, C. D. (1969). Enzymatic synthesis of (—)-kaurene and related diterpenes. *Methods Enzymol.* **15**, 481–490.

Wetmore, R. H., and Rier, J. P. (1963). Experimental induction of vascular tissues in callus of angiosperms. *Amer. J. Bot.* **50**, 418–430.

White, P. R. (1941). Plant tissue cultures. *Biol Rev.* **16**, 34–48.

White, R. P. (1946). Plant tissue culture. II. *Bot. Rev.* **12**, 521–529.

Whiteside, T. (1970a). "Defoliation." Ballantine Books, New York.

Whiteside, T. (1970b). Letter in the "Department of Amplification." *New Yorker* [*June* 20], p. 78.

Whitfield, P. R., and Spencer, D. (1968a). Buoyant density of tobacco and spinach chloroplast DNA. *Biochim. Biophys. Acta* **157**, 333–334.

Whitfield, P. R., and Spencer, D. (1968b). The biochemical and genetic autonomy of chloroplasts. *In* "Replication and Recombination of Genetic Material" (W. J. Peacock and R. D. Brock, eds.), pp. 74–86. Australian Acad. Sci., Canberra.

Whyte, R. O. (1948). History of research in vernalization. *In* "Vernalization and Photoperiodism" (A. E. Murneek and R. O. Whyte, eds.), pp. 1–38. Chronica Botanica, Waltham, Massachusetts.

Wiesner, J. (1892). "Die Elementarstructur und das Wachsthum der Lebenden Substanz." Alfred Hölder, Vienna.

Wightman, F., and Setterfield, G., eds. (1968). "Biochemistry and Physiology of Plant Growth Substances." Runge Press, Ottawa.

Wilkins, M. B., ed. (1969). "The Physiology of Plant Growth and Development." McGraw-Hill, New York.

Williams, C. M. (1969). Nervous hormonal communication in insect development. *Develop. Biol. Suppl.* **3**, 133–150.

Williams, G. R., and Novelli, G. D. (1968). Ribosome changes following illumination of dark grown plants. *Biochim. Biophys. Acta* **155**, 183–192.

Wilson, K. (1964). The growth of plant cell walls. *Int. Rev. Cytol.* **17**, 1–49.

Wimber, D., and Quastler, H. (1963). A C^{14} and H^{3}-thymidine double labeling technique in the study of cell proliferation in *Tradescantia* root tips. *Exp. Cell Res.* **30**, 8–22.

Winter, A. G. (1961). New Physiological and biological aspects in the interrelationships between higher plants. *Symp. Soc. Exp. Biol.* **15**, 229–244.

Wood, H. N. (1970). Revised identification of the chromophore of a cell division factor from crown gall tumor cells of *Vinca rosea* L. *Proc. Nat. Acad. Sci. U. S.* **67**, 1283–1287.

Wood, H. N., and Braun, A. C. (1967). The role of kinetin (6-furfurylamino-purine) in promoting division in cells of *Vinca rosea* L. *Ann. N. Y. Acad. Sci.* **144**, 244–250.

Wood, H. N., Braun, A. C., Brandes, H., and Kende, H. (1969). Studies on the distribution and properties of a new class of cell-division promoting substances from higher plant species. *Proc. Nat. Acad. Sci. U.S.* **62**, 349–356.

Woolhouse, H. W., ed. (1967). "Aspects of The Biology of Ageing." Academic Press, New York.

Worley, J. F. (1968). Intracellular injection of growth-regulating substances and their effect on the living protoplast of injected and adjacent cells. *In* "Biochemistry and Physiology of Plant Growth Substances" (F. Wightman and G. Setterfield, eds.), pp. 1635–1642. Runge Press, Ottawa.

Wright, S. T. C. (1969). An increase in the "Inhibitor-β" content of detached wheat leaves following a period of wilting. *Planta* **86**, 10–20.

Wright, S. T. C., and Hiron, R. W. P. (1970). The accumulation of ABA in plants during wilting and under stress conditions. *Seventh International Conference on Plant Growth Substances* No. 107, *December 7–12* (Abstr.). *Canberra, Australia.*

Xhaufflaire, A., and Gaspar, T. (1968). Les cytokinines. *Année Biol.* **7**, 39–87.

Yang, S. F. (1968). Biosynthesis of ethylene. *In* "Biochemistry and Physiology of Plant Growth Substances." (F. Wightman and G. Setterfield, eds.), pp. 1217–1228. Runge Press, Ottawa.

Yoshikami, D. (1970). Fluorescent transfer ribonucleic acids: Discovery and characterization. Ph.D. Thesis, Cornell University, Ithaca, New York.

Yunghans, H., and Jaffe, M. J. (1970). Phytochrome controlled adhesion of mung bean root tips to glass: A detailed characterization of the phenomena. *Physiol. Plant.* **23**, 1004–1016.

Zachau, H. G. (1969). Transfer ribonucleic acids. *Angew. Chem. Int. Ed.* **8**, 711–727.

Zachau, H. G., Dütting, D., Feldmann, H., Melchers, F., and Karau, W. (1966). Serine specific transfer ribonucleic acids. XIV. Comparison of nucleotide sequences and secondary structure models. *Cold Spring Harbor Symp. Quant. Biol.* **31**, 417–424.

Zelitch, I. (1969). Stomatal control. *Annu. Rev. Plant Physiol.* **20**, 329–350.

Zenk, M. H., and Nissl, D. (1968). Evidence against an allosteric effect of indole-3-acetic acid synthase. *Naturwissenschaften* **55**, 84–85.

Ziegler, H. (1970). Morphactins. *Endeavour* **29**, 112–116.

Zimmerman, P. W., and Hitchcock, A. E. (1933). Initiation and stimulation of adventitious roots caused by unsaturated hydrocarbon gases. *Contrib. Boyce Thompson Inst.* **5**, 351–359.

Zimmerman, P. W., and Wilcoxon, F. (1935). Several chemical growth substances which cause initiation of roots and other responses in plants. *Contrib. Boyce Thompson Inst. Res.* **7**, 209–229.

Zwar, J. A., and Brown, R. (1968). Distribution of labelled plant growth regulators within cells. *Nature* (*London*) **220**, 500–501.

Subject Index

D

E

F

G

J

K

L

M

Q

R